Mesay Kebede Duguma

Households' Food Insecurity and Coping Strategies in the Face of Vulnerability

Spektrum

Berlin Series on Society, Economy and Politics in Developing Countries

edited by

Prof. Dr. Dr. Dr. Ulrich van der Heyden
Prof. Dr. Volker Lühr
PD Dr. Ulrike Schultz
Prof. Dr. Manfred Schulz

Volume 110

LIT

Mesay Kebede Duguma

Households' Food Insecurity and Coping Strategies in the Face of Vulnerability

A gender perspective with empirical evidence from Meskan District, Southern Ethiopia

LIT

Published with the kind support of the German Academic Exchange Service / Deutscher Akademischer Austauschdienst (DAAD)

Bibliographic information published by the Deutsche Nationalbibliothek
The Deutsche Nationalbibliothek lists this publication in the Deutsche Nationalbibliografie; detailed bibliographic data are available in the Internet at http://dnb.d-nb.de.

ISBN 978-3-643-90608-3
Zugl.: Bonn, Univ., Diss., 2014

A catalogue record for this book is available from the British Library

Klosbachstr. 107
CH-8032 Zürich
Tel. +41 (0) 44-251 75 05
Fax +41 (0) 44-251 75 06
E-Mail: zuerich@lit-verlag.ch
http://www.lit-verlag.ch

LIT VERLAG Dr. W. Hopf
Berlin 2015
Fresnostr. 2
D-48159 Münster
Tel. +49 (0) 2 51-62 03 20
Fax +49 (0) 2 51-23 19 72
E-Mail: lit@lit-verlag.de
http://www.lit-verlag.de

Distribution:

In the UK: Global Book Marketing, e-mail: mo@centralbooks.com
In North America: International Specialized Book Services, e-mail: orders@isbs.com
In Germany: LIT Verlag Fresnostr. 2, D-48159 Münster
Tel. +49 (0) 2 51-620 32 22, Fax +49 (0) 2 51-922 60 99,
E-mail: vertrieb@lit-verlag.de

In Austria: Medienlogistik Pichler-ÖBZ, e-mail: mlo@medien-logistik.at
e-books are available at www.litwebshop.de

Acknowledgements

I express my heart-felt gratitude to my first supervisor, Prof. Dr. Britta Klagge for her unreserved scientific guidance and encouragement. Her constant support from University of Osnabrück to University of Bonn was very decisive to the completion of this thesis. I would like to thank Prof. Dr. Sabina Tröger, my second supervisor, for her willingness to evaluate my thesis. My appreciation also goes to Prof. Dr. Hans- Joachim Wenzel, Professor at the University of Osnabrück for his valuable comments on the first draft of this thesis. Special thanks to Dr. Katrin Klein-Hitpass and Dr. Benjamin Etzold for reading and translating the summary into German Language.

I am grateful to the German Academic Exchange Service (DAAD) for offering me a scholarship to study in Germany. I am also thankful to the Pool Frauenförderung, at the University of Osnabrück for financially supporting my field research in Ethiopia. I should not miss thanking Ms. Netsanet Mengistu, executive director of Progynist women Empowerment NGO for providing vehicles and rooms for undertaking focus group discussions at the Butajira branch office. My special thanks also go to the staff at Progynist for allowing me into their archives and assisting me to coordinate meetings with key informants.

I would like to extend my deepest gratitude to Prof. Dr. Tesfaye Tafesse, my academic father and mentor back in Addis Ababa University whose professional guidance, moral support and encouragement throughout my PhD years has been a tremendous input. I heartily thank my father, Kebede Duguma for his constant love and assistance throughout my study period. A special thanks to Dr. Girma Kelboro, a Post doc fellow at Center for Development Research (ZEF) for his sincere friendship and experience sharing during the last year of my study. My thanks also go to Daniel Kramer for his technical assistance. I acknowledge the following members of geography institute who helped me in one way or another: Pedro Campos Silva, Janina Wölfel, Jan Mörs, Irene Hillmer, Pia Sperber, Sebastian Reimer, Johanna Kramm and Julia Willers.

I am indebted to the Ethiopian community at Bonn Harvest Ethiopian Evangelical Church for their prayer and support during difficult times. Above all, I praise my God in the name of Jesus Christ. Without his loving care and protection, my academic years could have never bear fruits.

Dedicated to
My late mother Mihret Belayneh

Table of Contents

List of Tables

List of Figures

Glossary of Local Terms

Aba gorba	Blackleg
Aba sanga	Anthrax
Aba Wera	A male head
Belg	Short rains between February and May in Southern, north-eastern, and north central parts of Ethiopia
Birr	Local currency of Ethiopia
Chat	A shrub (Catha edulis) whose leaves, when chewed, provide a stimulant drug
Debo	A labour sharing arrangement in which a gathering of people are willing to provide hand
Dega	Cold temperate climate
Ekub	Group credit and savings scheme
Enset	The "false banana" tree (Ensete ventricosum) whose corn and stem provide a starch-based food
Ergiman	Spiritual curse
Fengil	Newcastle disease
Idir	Indigenous savings schemes to help families with money and funeral procedures
Kebele	The smallest administrative unit, rural or urban
Kiremt	The long rains between June and September
Meher	Main growing season, associated with the *kiremt* rains
Mekenajo	A means of resource sharing used in rural areas to acquire additional oxen mainly for ploughing
Shimagile	Wise old man
Tasa	local unit measurement of weight
Teff	A very fine-sized grain (Eragrostis *tef*) with a grass-like stem, unique to Ethiopia as staple
Timad	Local unit measurement of length
Woina dega	Mild temperate climate
Woreda	District
Yetelefa was	An arrangement in a group loan in which each member is a collateral for the entire group.
Zengada	Indigenous variety of Sorghum

List of Acronyms

ALDI	Agricultural Development-Led Industrialization
CARE	Cooperative for Assistance and Relief Everywhere
CBO	Community Based Organizations
CEDAW	Committee on the Elimination of Discrimination against Women
CSA	Central statistical Authority
DA	Development Agent
DFID	Department for International Development
DHS	Demographic and Health Survey
DPED	Department of Planning and Economic Development
EPDRF	Ethiopian Peoples' Democratic Revolutionary Front
FAD	Food Availability Decline
FAO	Food and Agricultural Organization
FDRE	Federal Democratic Republic of Ethiopia
FED	Food entitlement decline
FGD	Focus Group Discussion
FGM	Female Genital Mutilation
FHH	Female-Headed Households
FSP	Food Security Program
FSR	Farming System Research
GAD	Gender and Development
GDP	Gross Domestic Product
GTP	Growth and Transformation Plan
HABP	Households Asset Building Programme
HTP	Harmful Traditional Practice
ICCPR	International Convention on Civil and Political Rights
ICESCR	International Covenant on Economic, Social and Cultural Rights
ITK	Indigenous Technical Knowledge
MFI	Micro Finance institution
MHH	Male-Headed Households
MoARD	Ministry of Agriculture and Rural Development
NAP-GE	National Action Plan on Gender Equity
NGO	Non-Governmental Organization
NPEW	National Policy on Ethiopian Women

ODI	Overseas Development Institute
PASDEP	Plan for Accelerated and Sustained Development to End Poverty
PA	Peasant Association
PLA	Participatory Learning and Action
PRA	Participatory Research Appraisal
PSNP	Productive Safety Net Programme
RRA	Rapid Rural Appraisal
RUSACCO	Rural Saving and Credit Cooperatives
SDPRP	Ethiopian Sustainable Development and Poverty Reduction Programme
SLF	Sustainable Livelihood Framework
SNNPR	Southern Nations, Nationalities and Peoples' Region
UDHR	Universal Declaration of Human Rights
UN	United Nation
WAD	Women's Affairs Department
WAO	Women's Affairs Office
WAS	Women's Affaire Sector
WB	World Bank
WID	Women in Development

Summary

Studies dealing with the food insecurity and vulnerability of female-headed households in rural Ethiopia often emphasize their lack of access to livelihood resources, but often fail to deeply examine the interaction between access to resources and social institutions in explaining vulnerability. This study aims to gain an improved understanding of the gender dimension of vulnerability, food insecurity and coping strategies by placing formal and informal institutions and gender power relations at the centre of the investigation. The study proceeds from the general assumption that unequal gender power relations frame the vulnerability context of female-headed households, leading to greater food insecurity and weaker coping strategies as compared to their male counterparts.

The study employs the Sustainable Livelihood Framework (SLF) and Giddens' 'theory of structuration' as its conceptual framework. The SLF has provided methodological, organizational and analytical guidance for the empirical investigation in this thesis. The gender dimension of power relations at extra-household level is explored through the institution-agency relationships discussed in Giddens' "theory of structuration." Norms and customary practices represent the institutional aspect, while the actors are men and women heads of households, community-level leaders, development agents, district-level agriculture and gender experts and *kebele* officials.

A mixed research approach was employed for the empirical investigation carried out in four rural administrative *kebeles* in Meskan district of southern Ethiopia. The primary data was collected in two rounds (between 2 January and 3 May 2011 and in May 2012). Qualitative information was gathered on a wide range of institutional issues through focus group discussions and key informant interviews, case stories and participant observation. A total of 9 focus group discussions and 17 key informant and expert interviews were conducted with local actors during the four months of fieldwork. Quantitative information was gathered through a household survey conducted in 2011 involving a total of 372 female- and male-headed households selected using a simple random sampling technique.

The findings show that in the traditional marriage and family institutions women are denied the opportunity to exercise their right to education, a key factor in accessing all other economic, social and political resources for building a secure livelihood. Moreover, social norms and harmful traditional practices erode women's confidence and self-esteem, thus increasing their economic dependence on men and their level of insecurity when confronted with livelihood shocks and stresses. The prevailing gender roles reinforce the exclusion of women from important decision-making processes. Institutional pluralism in the existing customary judiciary system undermines women's rights during disputes over resources. The study thus confirms the hypothesis that in the face of institutional pluralism, local informal institutions shape the vulnerability context of female-headed households in Meskan district.

The study further reveals that a greater proportion of female-headed households are seasonally food insecure, as compared to MHHs. On average, female-headed households own smaller plots of land and have limited control over it. Nearly 52 percent of female-headed households, against 30 percent of MHHs, do not possess oxen. They suffer from shortage of labour and time to engage in various on-farm and off-farm agricultural activities, have fewer contacts with agricultural experts and benefit less from agricultural extension services. Nearly 54 percent of the respondents in female-headed households, as against only 34 percent in male-headed households, did not apply chemical fertilizer in the production year 2010. Poor access to potable water and water infrastructures imposes unique challenges on female-headed households. Female-headed households have restricted powers of decision-making in livelihood spheres that are by tradition assumed to be "male domains." The findings support the hypothesis that female-headed households face greater food insecurity than male-headed households, due to their lack of access to productive resources and decision-making powers.

The study identifies illness of a family member, death of domestic animals, crop failure and shortage of income as some of the most serious forms of livelihood shocks and stresses experienced by farming households. The survey also reveals that in the face of weakening traditional coping mechanisms, a large proportion of both female- and male-headed households use coping strategies such as dependence on market exchange of livestock, shifts in planting decisions, and diversification into off-farm activities to survive frequently occurring yet unprecedented crises. Yet, women farmers face cultural barriers to effectively employ these coping strategies. They are not able on their own to exchange cattle on the market. Crop diversification is less practiced on farm lands owned by female heads. Moreover, women largely engage in less lucrative petty trading and often face restricted entry to the wage labour market as compared to their male counterparts. The study concludes that female-headed households have weaker coping strategies against livelihood shocks and stresses as compared to their male counterparts.

The findings of the study with regard to the role of external interventions in filling the gender gap in food security and empowering women in the study *kebeles* revealed certain general and gender-specific constraints. On the one hand, interventions such as the Productive Safety Net Programme (PSNP) that are widely implemented in the study area, have helped to decrease nutrition-related mortality, prevented asset depletion and brought about some attitudinal changes. On the other hand, lack of awareness among farmers, and perception gaps between officials and farmers are identified as some of the barriers to the smooth running of these programmes. From a gender perspective, some of the components of the programme that targeted women have overlooked the problems connected with the motivation and resource bases of FHHs. The public works dimension of the PSNP has inadequately addressed issues of water, energy and road infrastructures which are major challenges for rural women in general and FHHs in particular. However, an NGO intervention in Beresa *kebele* has played a key role in creating awareness of women's rights and gender equality among the study population. Yet, the existing situation is still far from bringing changes in terms of eliminating discriminatory perceptions and practices against women in all of the study *kebeles*. The study shows that there is no effective women's policy-implementing body. Lack of personnel and financial resources, and inability to extend to grassroots level, are identified as the major gaps.

In all areas examined, the study confirms its initial presumption that unequal gender power relations frame the vulnerability context of female-headed households, leading to greater food insecurity and weaker coping strategies as compared to their male counterparts.

Zusammenfassung

Viele Studien, die sich mit Ernährungsunsicherheit und Vulnerabilität frauengeführter Haushalte im ländlichen Äthiopien beschäftigen, betonen oftmals den fehlenden Zugang zu Ressourcen zur Existenzsicherung dieser Haushalte. Häufig lassen sie aber bei der Erklärung von Vulnerabilität eine tiefergehende Untersuchung des Zusammenhangs zwischen dem Zugang zu diesen Ressourcen und sozialen Institutionen außer Acht. Ziel dieser Studie ist es, ein verbessertes Verständnis der Gender-Dimension von Verwundbarkeit, Ernährungsunsicherheit und Bewältigungsstrategien angesichts externer Schocks zu gewinnen. In diesem Kontext werden insbesondere formale und informelle Institutionen sowie geschlechtsspezifische Machtverhältnisse in den Mittelpunkt der Untersuchung gestellt. Die Studie geht von der allgemeinen Annahme aus, dass ungleiche geschlechtsspezifische Machtverhältnisse die Verwundbarkeit frauengeführter Haushalte prägen. Dies führt zu einer höheren Nahrungsmittelunsicherheit und schwächeren Bewältigungsmechanismen im Vergleich zu Haushalten mit männlichem Vorstand.

Die Studie nutzt das Sustainable Livelihood Framework (SLF) und Giddens Strukturationstheorie als konzeptionelle Grundlage. Das SLF bietet einen methodischen, organisatorischen und analytischen Rahmen für die empirischen Untersuchungen in dieser Arbeit. Die Gender-Dimension von Machtverhältnissen jenseits der Haushaltsebene wird mit Hilfe der Struktur -Handlungs-Relation gemäß Giddens Strukturationstheorie untersucht. Dementsprechend werden geschlechtsspezifische Machtverhältnisse jenseits der Haushaltsebene durch die Beziehungen zwischen Institutionen und Akteuren erforscht. Normen und traditionelle Gepflogenheiten stellen die institutionellen Aspekte dar, während die weiblichen und männlichen Haushaltsvorstände, Führungskräfte der Gemeindeebene, Entwicklungsberater, Landwirtschafts- und Gender-Experten auf der Distriktebene sowie *Kebele*-Beamte die im Zentrum stehenden Akteure dieser Studie sind.

Die empirische Untersuchung in vier ländlichen *kebeles* im Meskan Distrikt im Süden Äthiopiens kombiniert verschiedene Methoden. Die Primärdaten wurden in zwei Erhebungsphasen (2. Januar bis 3. Mai 2011 und Mai 2012) erhoben. Qualitative Daten zu einer Vielzahl von institutionellen Aspekten wurden durch Fokusgruppendiskussionen, Interviews, Einzelfallstudien und teilnehmende Beobachtung gewonnen. Insgesamt neun Fokusgruppendiskussionen und 17 Experteninterviews wurden in den vier Monaten der Feldarbeit durchgeführt. Quantitative Daten wurden durch eine Haushaltsbefragung in 2011 gewonnen. Insgesamt wurden 372 weibliche und männliche Haushaltsvorstände mit Hilfe einer einfachen Zufallsstichprobe ausgewählt und befragt.

Die Ergebnisse der Studie zeigen, dass Frauen im Rahmen der traditionellen Institutionen Ehe und Familie ihr Recht auf Bildung verweigert wird. Bildung ist allerdings ein Schlüsselfaktor für den Zugriff auf alle anderen wirtschaftlichen, sozialen und politischen Ressourcen zur Existenzsicherung. Darüber hinaus untergraben herrschende soziale Normen und traditionelle Praktiken das Selbstvertrauen und Selbstwertgefühl vieler Frauen. Damit steigt ihre wirtschaftliche Abhängigkeit von Männern und der Grad ihrer Unsicherheit, vor allem in besonderen Belastungssituationen, die ihre Existenz bedrohen. Darüber hinaus führen vorherrschende Geschlechterrollen zum Ausschluss von Frauen aus wichtigen Entscheidungsprozessen. Die existierende Pluralität im Rechtssystem untergräbt zudem die Rechte der Frauen, insbesondere bei der Verteilung von Ressourcen. Folglich bestätigt die Studie die Hypothese, dass angesichts des institutionellen Pluralismus lokale informelle Institutionen die Verwundbarkeit frauengeführter Haushalte im Meskan Distrikt stark beeinflussen.

Die empirischen Ergebnisse der Studie zeigen zudem, dass ein größerer Anteil von Haushalten mit weiblichem Vorstand von temporärer Ernährungsunsicherheit betroffen ist als Haushalte mit männlichem Vorstand. Im Mittel besitzen die von Frauen geführten Haushalte kleinere Grundstücke und zudem ist ihre Kontrolle über diese Grundstücke begrenzt. Fast 52 Prozent der von Frauen geführten Haushalte besitzen gegenüber 30 Prozent der von Männern geführten Haushalte keine Ochsen. Sie haben weder die Arbeitskraft noch die Zeit, um innerhalb und außerhalb der eigentlichen Landwirtschaft zu arbeiten und ein zusätzliches Einkommen zu generieren. Darüber hinaus haben sie weniger Kontakte zu Agrarexperten und profitieren daher auch weniger von landwirtschaftlicher Beratung. Fast 54 Prozent der Befragten in von Frauen geführten Haushalten gegenüber nur 34 Prozent in von Männern geführten Haushalten nutzten im Produktionsjahr 2010 keinen chemischen Dünger. Unzureichender Zugang zu Trinkwasser und Wasserinfrastruktur stellen außerdem hohe Herausforderungen an die weiblichen Haushaltsvorstände dar. Frauen haben einer geringeren Einfluss auf Entscheidungsfindungsprozesse im Bereich Überlebensstrategien, die traditionell als ‚männliche Domänen' angesehen werden. Die Forschungsergebnisse unterstützen die These, dass Frauen gerade aufgrund ihres mangelnden Zugangs zu produktiven Ressourcen und ihrer schwächeren Machtposition im größeren Maße von Hunger und Mangelernährung betroffen sind.

Die Studie hat ergeben, dass Krankheit eines Familienmitglieds, der Tod von Nutztieren, Ernteausfälle sowie andere Verluste hinsichtlich des Einkommens zu den schwerwiegendsten Problemen zählen, welche die Existenz Landwirtschaft betreibender Haushalte gefährden können. Darüber hinaus wurde herausgearbeitet, dass häufiger auftretende schwere Krisen immer seltener durch traditionelle Anpassungsmechanismen gelöst werden. Immer mehr Haushalte, seien es männer- oder frauengeführte Haushalte, verwenden Strategien wie den marktbasierten Viehhandel, veränderte Anbaumethoden oder eine Diversifizierung der Existenzsicherung durch Aktivitäten außerhalb der Landwirtschaft. Frauengeführte Haushalte sehen sich aber mit verschiedenen kulturellen Barrieren konfrontiert, die sie daran hindern, diese alternativen Strategien effektiv anzuwenden. Hierzu gehört, dass es ihnen nicht möglich ist, auf sich allein gestellt auf dem Markt Handel mit Vieh zu betreiben. Außerdem werden diversifizierte Anbaumethoden seltener auf Anbauflächen angewandt, die zu einem frauengeführten Haushalt gehören. Zudem ist der von Frauen betriebene Handel oft kein für sie ausreichend Gewinnbringendes Geschäft. Darüber hinaus haben Frauen einen erschwerten Zugang zum Arbeitsmarkt im Vergleich zu Männern. Die Studie kommt zu dem Schluss, dass frauengeführten Haushalten weniger Möglichkeiten zur Verfügung stehen, ihre Existenz gegenüber Bedrohungen und Stressoren zu schützen.

Die Forschungsergebnisse zeigen bestimmte allgemeine und genderspezifische Hemmnisse bei der Frage, welche Rolle externe Interventionen bei der Überwindung des gender gaps hinsichtlich Ernährungssicherheit und der Rechte der Frauen in den Forschungsregionen spielen. Einerseits haben Interventionen wie das Productive Safety Net-Programm (PSNP) dazu beigetragen, die ernährungsbedingte Sterblichkeit zu verringern und den Verlust von Produktionsgütern zu verhindern. Darüber hinaus konnten einige Veränderungen hinsichtlich der Einstellung gegenüber bestimmten Aspekten erreicht werden. Andererseits wurde der reibungslose Ablauf der Programme durch mangelndes Bewusstsein der Bauern sowie durch Einstellungsdifferenzen zwischen Beamten und Bauern gestört. Aus einer Gender-Perspektive wurden in einigen speziell auf Frauen ausgerichteten Elementen des Programms die spezifischen Probleme der von Frauen geführten Haushalte bezüglich Motivation und Ressourcenbasis übersehen. Die öffentlichen Bauvorhaben des PSNP haben Fragen der Wasser-, Energie- und Straßeninfrastruktur nur unzureichend behandelt. Diese stellen aber Frauen in ländlichen Gebieten und insbesondere weibliche Haushaltsvorstände vor große

Herausforderungen. Nichtsdestotrotz hat die Maßnahme einer NGO in Beresa *Kebele* bei der Sensibilisierung der Bevölkerung für die Rechte von Frauen und die Gleichstellung der Geschlechter eine Schlüsselrolle gespielt. Dennoch ist die derzeitige Situation in allen Forschungsregionen noch weit davon entfernt, diskriminierende Wahrnehmungen und Praktiken gegenüber Frauen zu beseitigen. Die Studie zeigt, dass es keine wirksame Institution gibt, welche Strategien zur Wahrung der Rechte und Bedürfnisse der Frauen implementiert. Ein Mangel an personellen und finanziellen Ressourcen sowie die Unfähigkeit, die Politiken auf lokaler Ebene zu implementieren, wurden als wesentliche Hindernisse indentifiziert.

In allen untersuchten Bereichen bestätigt die Studie die anfängliche Vermutung, dass ungleiche geschlechtsspezifische Machtverhältnisse die Verwundbarkeit frauengeführter Haushalte wesentlich prägen. Dies führt unter anderem dazu, dass Frauen häufiger mit Ernährungsunsicherheit konfrontiert sind und sie über weniger Bewältigungsstrategien gegenüber Krisen verfügen, als dies bei Männern der Fall ist.

1 Introduction

1.1 Background

Agriculture plays a crucial role in Ethiopia's national economy. It contributes to nearly 40 percent of the country's gross domestic product (GDP) and provides a livelihood for 80-85 percent of the country's population.[1] It supplies raw materials to the country's agro-industrial sector, and is the main source of national export earnings. Agriculture also plays a key role in the provision of food for the population and hence is the prime contributing sector for the country's food security. However, this sector is characterized by a small-scale subsistence farming system. More than 95 percent of the farming is dependent on seasonal rain (Alemayehu 2006, 5).

Agriculture and rural enterprises in developing countries significantly rely on the contribution of women as farmers, workers and entrepreneurs (FAO 2011a). Women in Sub-Saharan Africa account for up to 50 percent of the agricultural labour force (FAO 2011b, 4). Women are not just agricultural labourers; they are also decision makers as well as managers of their own households (Saito and Weidemann 1990). Similarly, rural women in Ethiopia significantly contribute to the agricultural workforce, either as family members or as household heads (Aregu *et al.* 2011). A study by Frank (1999) on the role of gender in the Amhara region finds that women carry out tasks such as land preparation, weeding, harvesting, threshing and storing, in addition to their primary role in household gardening. In the livestock production sector, women are often responsible for herding, tending sick animals, watering, barn cleaning, milking and milk processing.

Yet, despite the enormous contribution of rural women in the agricultural sector, ample empirical evidence from different regions of the world confirms that women enjoy fewer privileges in respect of acquiring access to agricultural productive resources as compared to their male counterparts (Lado 1992; FAO 2011a; Agarwal 1988; Carney and Watts 1991; Lastarria-Cornhiel 1997). Women farmers across Africa share common barriers that put restrictions on their productivity, thereby affecting their overall well-being at the family, community and national levels. They enjoy less access to information, technology, land, inputs, markets and credit as compared to their male counterparts. As primary care-takers of children, they have less time and are often illiterate (Saito and Weidemann 1990, II).

In a patriarchal society like Ethiopia, the contribution of rural women to the economic and social development of the country often receives little recognition, creating significant gender disparities in access to resources and decision-making power and the sharing of development benefits (Yigremew 2004; Flintan *et al.* 2008; Frank 1999). Various empirical studies in Ethiopia reveal notable gender asymmetries in *de facto* access to and control over land (Crewett *et al.* 2008; Fafchamps and Quisumbing 2000). Moreover, although female-headed households (FHHs) enjoy access to land, which is often much smaller in size than that of male-headed households (MHHs), other productive resources such as labour, oxen, and credit are usually less accessible to them. The study by Bigsten *et al.* (2002) finds that among the causes of rural women's poverty is inadequate access to resources in general and to rural land in particular.[2] As a result, women in general and female-headed households in particular form categories with chronic food insecurity (Clay *et al.* 1999; Ayele *et al.* 2003). Citing the report of the

[1] According to the 2013 World Bank report, the total population of Ethiopia is estimated to be around 91 million in the year 2012. [data.worldbank.org/country/ethiopia accessed on 20/11/2013]

[2] Bigsten *et al.* 2002 examined the causes of falling into and escaping from poverty by looking at variables such as land ownership, education, type of crops planted, occupations in urban areas and dependency ratios. They analysed panel data covering the period 1994–1997 in selected rural and urban areas of Ethiopia.

New Coalition for Food Security (2003), Yigremew (2006, 20) states that a significant proportion of the 5 million most chronically food insecure people in the country are women, specifically identifying them as female-headed households. The Sustainable Development and Poverty Reduction Programme (SDPRP) for the year 1999/2000, quoted in Yigremew (2006, 19), also revealed that female-headed households face 8.9 percent higher probability of being poor as compared to male-headed households.

In an era where eliminating hunger and extreme poverty is closely tied to promoting gender equality and empowerment of women in the agricultural sector (FAO 2011a), neglecting the role and contribution of women and the failure to solve their underlying structural problems is extremely counter-productive. The negative effects of such an oversight are tangibly felt due to the damage it inflicts on the smooth running of societal life in which poverty, food insecurity and increasing vulnerability are rampant. In cognizance of this, several policies and strategies have been designed to gear development towards improving the condition of women in general and of rural women in particular. The National Policy on Ethiopian Women (1993) and the Food Security Strategy (2002)[3] are among the most important initiatives made by the government towards the fulfilment of this goal. On top of that, major changes in legislation were made in the 1995 Federal Constitution guaranteeing that women have equal rights with men. In terms of major development programmes, the Ethiopian Sustainable Development and Poverty Reduction Programme (SDPRP, 2000-2004/5) considers gender as a cross-cutting issue. Similarly, the two subsequent major government development programmes, namely, A Plan for Accelerated and Sustained Development to End Poverty (PASDEP, 2005/6-2010/11) and the Growth and Transformation Plan (GTP, 2010/11-2014/15) treat gender as a crucial area of concern in respect of development.

Yet, despite the above stated policy provisions, rural women face a number of institutional barriers, including cultural and social constraints in the form of gender-biased customary practices, conventions, stereotypes and misconceptions. These barriers prevent them from fully enjoying access to and control over productive resources such as land, labour and credit, and agricultural technologies (Ogato *et al.* 2009). As a result, bringing about a viable change in terms of solving these underlying structural problems that affect a large segment of women in general and female-headed households in particular still remains a formidable challenge. Many of the practical as well as the strategic needs of women in the rural areas still remain a huge concern.

1.2 Statement of the problem

Environmental changes, such as increasing degradation of resources, increase in the frequency of droughts and changes in rainfall patterns have overwhelmed the subsistence farming system, threatening food security and the traditional coping mechanisms employed by farm households in the study area. The impact of such changes, however, is felt most heavily by resource-poor and socially marginalized groups such as FHHs. Taking into consideration the existing gender inequality and its role in aggravating poverty, the government's poverty-reduction oriented development programmes that have been implemented in recent decades have aimed at reconstructing the agricultural sector. Various policy and legislative measures have been taken to integrate rural women in general and women farmers in particular, in the agricultural development process. These measures have been followed by certain transformations in the social, economic and legal settings which shape the landscape against which gender-based power relations can be understood.

3 See chapter three for a review of the National Policy on Ethiopian Women and the Food Security Strategy under the umbrella of the various development programmes of the country.

The gender dimension of poverty and food insecurity is critical to Ethiopia's development, and thus requires a thorough understanding of women's roles, their decision making power and the level of recognition of their role in the overall social and economic development of the country. Female-headed households are also among the groups that need special attention. There are plenty of studies on gender roles, gender power relations and decision-making in Ethiopia, more now than ever before. Yet, much focus is on intra-household gender power relations and decision making in male-headed households instead of female-headed households. In this regard, even though many studies point out the lack of access to resources of female-headed households as compared to their male counterparts (Dejene 1994; Etenesh 1999; Ali 2000; Kumar and Quisumbing 2012, Ogato *et al.* 2009), there are few studies that deal in depth with such disadvantages through a social, cultural and institutional lens. Thus, most of the studies pay less attention to examining the role of gender power relations and the obstacles to decision-making at extra household level in dealing with the factors that perpetuate vulnerability and food insecurity in female-headed households.

Against the above backdrop, this thesis seeks to explore the vulnerability context of men and women farmers, and the factors leading to the gender gap in food security and in the capacity to cope with livelihood shocks and stresses, with due consideration of the gender power relations that play out beyond household level. This requires a thorough investigation of the institutional set up and the social processes that shape the roles, access to resources, preferences and decision-making powers of men and women farmers. A systematic generation of area-specific knowledge through an in-depth investigation of the views and opinions of the very people who are involved in the process is an appropriate research approach to the acquisition of qualitative information on the subject. The study recognizes the above knowledge gap and examines the vulnerability, food insecurity and coping strategies of households through a close investigation of the interplay between formal and informal institutions and farmer's livelihood decisions.

Meskan district is a priority district for investigation in this thesis for two major reasons. Firstly, it is one of the districts in the Southern Nations, Nationalities and Peoples' Region (SNNPR) with a high prevalence of chronic food insecurity (MoARD, 2006). Secondly, gender based inequalities in the district are not only pervasive but also strongly backed up by patriarchal perceptions, customary practices, norms and values (Progynist 2002). Therefore, it presents an ideal research site for the empirical investigation that this thesis opts for.

1.3 Research hypotheses

Natural factors are significant triggers of food insecurity in Ethiopia, yet lately much emphasis has been given to the factors that increase the vulnerability of households to various livelihood shocks. It is stated in Moser (1998, 3) that *"analysing vulnerability involves identifying not only the threat but also the "resilience" responsiveness in exploiting opportunities, and in resisting or recovering from the negative effects of changing environment. The means of resistance are the assets and entitlements that individuals, households, or communities can mobilize and manage in the face of hardship."* From a gender perspective, abundant studies that focus on food security and coping strategies show the existence of a systematic link between gender and access to resources and options in livelihood strategies. Widespread patriarchal perceptions, customary practices and gender norms and values work in favour of men and against women. These widely acknowledged norms and perceptions impinge on women's role and mode of life in their community.[4]

[4] For review of literature on the gender dimension food insecurity and vulnerability, see 2.2

In light of the above, the research started off with the general hypothesis that in the face of common external livelihood shocks and stresses, unequal gender power relations frame the vulnerability context of female-headed households, leading to greater food insecurity and weaker coping strategies, as compared to their male counterparts.

The above hypothesis is further broken down in to the following three specific hypotheses that are dealt with separately in three different chapters (7, 8 and 9 respectively).

1. In the face of institutional pluralism, informal institutions largely shape the vulnerability context of female-headed households.
2. With less access to resources and decision-making powers, FHHs face greater food insecurity than their male counterparts.
3. As compared to MHHs, FHHs have weaker coping strategies against livelihood shocks and stresses.

1.4 Objectives

The overarching aim of the study is to gain an improved understanding of the gender dimension of vulnerability, food insecurity and coping strategy among female and male-headed households in Meskan district of southern Ethiopia.

The study also has the following specific objectives:

(a) To examine the vulnerability context in female- and male-headed households.

(b) To explore the determinant factors that lead to food insecurity in female- and male-headed households.

(c) To investigate the type of coping strategies against livelihood shocks employed in female- and male-headed households.

(d) To assess the role of food security programmes and gender policies in empowering women and addressing the gender gap in food security in Meskan district.

Reducing poverty and ensuring households' food security, both at the regional and the national level, continues to be one of the challenges Ethiopia faces to date. It is the author's belief that the diversity of the agro-ecological and livelihood systems in the country add another component to this challenge. Thus, effective intervention demands a policy recommendation that emanates from area-specific in-depth understanding of the actual social and economic realities on the ground. To achieve this, the study primarily adopts the "sustainable livelihoods framework" and "the theory of structuration" as the two complementing analytical and theoretical frameworks guiding the investigation. Both qualitative and quantitative methods are employed in which qualitative data is predominantly used to enrich the quantitative information gathered at household level. On top of that, the gender dimension is captured by using female- and male-headed households as the primary proxy unit of analysis.

1.5 Outline of the thesis

Following this introductory chapter which describes the problem addressed in the research, the second chapter presents the literature review and theoretical aspects of the thesis. It starts with introducing key terms and concepts followed by a literature review on gender and vulnerability with particular emphasis on Ethiopia. Then the chapter briefly touches up on the historical development of rural development discourses and gender studies. This is followed by a discussion of the theoretical frameworks adopted in this thesis. Chapter three presents an appraisal of rural development programmes and gender policies in Ethiopia, with a special focus on food security and gender policies. This chapter gives an overview of the policies of the national government in respect of ensuring food security and gender equality, and provides the basis for the empirical analysis presented later on in chapter ten. The fourth chapter describes the study district. Chapter five presents the research methods employed in the study. The subsequent five chapters (chapters six, seven, eight, nine and ten) deal with the empirical findings and discussions. Chapter six describes the demographic and socio-economic character of the study population. Chapter seven presents the findings with regard to vulnerability contexts and trends. This chapter provides the foundation for the analysis made in the subsequent chapters (chapters eight, nine and ten). Chapter eight focuses on the determinants of household food insecurity. Chapter nine presents the study findings on coping strategies that are employed by households against livelihood shocks and stresses. Chapter ten discusses external interventions. Here, the local impact of policy interventions with a particular focus on food security and gender dimensions is assessed and gaps are identified. The concluding chapter includes a discussion of possible implications for future policies.

2 Literature review and theoretical considerations

This chapter is divided into seven major sub-topics. It begins by introducing key terms and concepts followed by a brief review of gender and livelihood studies in Africa with a particular emphasis on Ethiopia. In the third sub-section, a review of rural development approaches and gender studies is presented. A historical approach is employed to show how rural development discourses and gender theories gradually grew up under the umbrella of wider development paradigms. It is shown that both fields laid the ground for one of the currently popular analytical tools in livelihood studies, the Sustainable Livelihoods Framework (SLF). The fourth and the fifth sections deal with the SLF and the "theory of structuration" which formed the conceptual backdrop of the thesis. The sixth sub-section discusses their link with the notion of vulnerability. The last section presents the analytical framework used for data analysis.

2.1 Introducing key terms and concepts

In order to facilitate a contextual understanding of certain terms used throughout this thesis, the following section discusses some of the key terms used in this thesis by showing their contextual relevance to the study. These include: gender and gender power relations, informal institutions, female-headed households, vulnerability, food (in)security and coping strategy. Emphasis is given to the conceptual linkage among the last three.

Gender and gender power relations: The term gender can be understood as "*the rules, norms, customs and practices by which biological differences between males and females are translated into socially constructed differences between men and women and boys and girls*" (Kabeer 2003, 2; IIRR-Ethiopia 2000).

Gender power relations are often described as a complex concept. As Agarwal (1997, 1-2) puts it: "*the complexity arises not least from the fact that gender relations (like all social relations) embody both the material and the ideological. They are revealed not only in the division of labour and resources between women and men, but also in ideas and representations ascribing to women and men different abilities, attitudes, desires, personality traits, behaviour patterns, and so on. Gender power relations are both constituted by and help constitute these practices and ideologies, in interaction with other structures of social hierarchy such as class, caste and race. They are neither uniform across societies nor are they historically static.*" This explanation supports the definition of gender as a socially constructed notion rather than a biologically determined characteristic.

Informal institutions: Informal institutions generally include non-officially enforced cultural norms, mechanisms, beliefs and social networks that are usually communicated through verbal channels (Helmke & Levitsky 2004). These are constituted in every community and serve as mechanisms by which interactions between people and their environment are governed. They are deeply entrenched in people's culture and traditions, and are passed from one generation to another. They have a number of functions in societal life, including protecting local traditions, preserving cultural and natural heritage, managing the environment and natural resources and mediating conflicts over the use of natural resources (Long 1997).

Female-headed households: the most common way in which women appear in poverty assessments is in the guise of female-headed households. Reviewing some of the World Bank's poverty assessment reports in Sub-Saharan Africa, starting from the 1980s, Whitehead and Lockwood (1999) assert that most gender-based poverty assessments use female-headed households as proxy units of analysis. However, a number of studies of the relationship between female headship and poverty have come up with controversial outcomes. Some show a positive association (Buvinic and Gupta 1997) and suggest the use of household head as a proxy unit of poverty analysis and development intervention. Others, such as the one by Quisumbing *et al.* (1995), which uses household survey data sets from 10 developing countries, have questioned the general application of the above suggestion. Thus, led to the conclusion that differences between poor female- and male-headed households are hardly significant, the studies suggest further disaggregation of female-headed household units across other variables. Such contradictory conclusions suggest that female-headed households need to be treated as heterogeneous groups rather than homogeneous ones (Quisumbing *et al.* 1995).

Female heads of households in Ethiopia are certainly a heterogeneous group, including women who are divorced, widowed, unmarried, abandoned, etc. In general a female-headed household is understood as "*a unit where an adult woman (usually with children) resides without a male partner*" (Ali 2000 cited in Yigremew 2001, 2). These households are further distinguished as *de jure* and *de facto*. In the Ethiopian rural context, a *de jure* female-headed household is defined as "*a household where the land is owned and managed by a woman, such as in families headed by widows or by single or divorced women*," while a *de facto* female-headed household is a household where "*a woman is responsible for all aspects of managing the household and farm due to the absence of a husband*" (Addis 2000, 15-16).

Female headship is increasingly becoming a common phenomenon in Africa (Lado 1992). Various studies made in different African countries confirm this, citing men's migration, among many other factors, as a cause of increasing female headship (World Bank 1989; Skapa 1988 and Spring 1985 cited in Saito and Weidemann 1990). In the case of Ethiopia, the Ethiopian Demographic and Health Survey (DHS) for the year 2000 indicated 21 percent female-headed households for rural areas and 35 percent for urban areas (CSA and ORC Macro 2001, 7). A more disaggregated report for the same year indicated that 20 percent of the rural households were *de jure* female-headed households, while *de facto* female-headed households constitute 10-15 percent (Addis 2000, 15-16). In general, female headship is assumed to be the outcome of the long civil war and the resulting high internal and external displacement on the one hand, and on the other hand a result of the crippling famines and resettlement schemes in the country (Dessalegn 1994, 5; Zenebework and Yared 2000, 30).

Vulnerability: The concept of vulnerability is one of the most frequently applied notions in livelihood research and its conceptualization and usage covers a wide range. As noted in Watts and Bohle (1993, 46), it has been discussed in ecological terms, in relation to political economy and class structure and as a reflection of social relations including ethnicity, caste, generation and gender.

Chambers (1989, 1) explains "*Vulnerability is not lack or want, but defenselessness, insecurity and exposure to risks and stress*". Chambers distinguishes two faces of vulnerability; an external side of risks, shocks and stress to which an individual or household is subject, and an internal side which is defenselessness, meaning lack of means to cope without damaging loss. Here, loss can take many forms including becoming or being physically weaker, economically impoverished, socially dependent, humiliated or psychologically harmed. The internal dimension of vulnerability is also emphasized in Wisner *et al.* (1994) where vulnerability deals with one's ability to cope with, resist, and recover from the impact of a natural hazard.

A more refined use of the concept of vulnerability is found in the food insecurity, or famine, and natural hazards literature. For instance, Watts and Bohle (1993) contributed to the theoretical understanding of the term in their article "*Hunger, famine and the space of vulnerability*", in which vulnerability is described as a multi-layered and multi-dimensional social space defined by the determinate political, economic and institutional capabilities of people in specific places at specific times (Watts & Bohle 1993, 118). Vulnerability is also explained through three basic coordinates. These are: the risk of exposure to crises, stress and shocks; the risk of inadequate capacities to cope with stress, crises and shocks; and the risk of severe consequences of, and the attendant risk of slow or limited recovery from, crises and shocks. In other words the above elaboration can be substituted by three key terms: exposure, capacity and potentiality (Bohle *et al.* 1994, 38).

It is very often argued that a household can experience vulnerability without actually being poor (Chambers 1989; Moser 1996; Razivi 1997; Wisner *et al.* 1994). Bohle *et al.* (1994, 38) argue that not all who are assumed to be poor carry a risk of equal vulnerability to hunger. Thus, not all who are considered poor at a given instance are equally vulnerable and even among those groups that are classified as vulnerable, a certain degree of variation is inevitable. Some are highly vulnerable and highly sensitive, while others are exposed in a less blatant manner. The most vulnerable individuals, groups, classes and regions are those most exposed to distress, those who possess the most limited coping capability, who suffer the most from crisis impact and who are endowed with the most restricted capacity for recovery. Thus, they further stress "*the prescriptive and normative response to vulnerability is to reduce exposure, enhance coping capacity, strengthen recovery potential and bolster damage control (that is minimize destructive consequences) via private and public means*" (Bohle *et al.* 1994, 38).

Generally, the applicability of the vulnerability concept in the analysis of food security and coping strategies is well documented. For instance the DFID (2002) document describes vulnerability in relation to food insecurity as the presence of factors that place people at risk of becoming food insecure or malnourished, including those that affect their ability to cope. Holmes and Jones (2009) similarly emphasize the importance of dealing with vulnerability in the analysis of food security. This is because a prior understanding of the vulnerability situation in food-insecurity-prone areas helps to tackle the negative coping strategies that undermine livelihoods, thus providing a basis for future food security.

Food (in)security: Food security has been defined in a number of ways. Various studies on Ethiopia associate food insecurity with access to resources, apart from the climatic and environmental causes. For instance, Degefa (2005, 10) describes a food secure household as a household where the livelihood activities are sufficient to meet the household's food requirements and other basic needs, either through the household's own production, i.e. crop cultivation and/or livestock rearing, or through having other opportunities to run non-farm ventures or to work for somebody else. Conversely, food insecurity refers to a situation where a household is not capable of adequately feeding its members from either its own production or market purchase. Moreover, in describing food security in the context of subsistence farming, Getachew (1995) defines food security as the ability to establish access to productive resources such as land, livestock, agricultural inputs and family labour in order to produce food or cash.

Complementing the above definitions, an important insight is found in Devereux (2000, 1) where he points out the vulnerability aspect of food insecurity. He argues that food insecurity incorporates low food intake, variable access to food, and vulnerability – resulting from a livelihood strategy that generates adequate food in good times but is not resilient against shocks. Vulnerability is therefore closely linked to asset ownership. According to Moser's (1998, 3) assertion, "*the more assets people have the less vulnerable they are, and the greater the erosion of people's assets, the greater their insecurity*." Thus, access to and control over assets is a crucial determining factor of livelihood outcomes, primarily because control over resources enables farm households to pursue alternative livelihood activities which lead to households' food security. Resources are also a source of resilience and a buffer at times of exposure to shocks and stresses, and are thus a substantial part of the coping strategies used by poor households.

Discussing famine in the Ethiopian context, Mesfin (1984) argues that a vulnerable agricultural population must first have been made vulnerable to famine by socio-economic and political forces, prior to the damage caused by any natural factor that leads to famine. Similarly, the 2009 MoARD document emphasizes that while droughts and other disasters are significant triggers of food insecurity in Ethiopia, more important are the factors which create and/or increase vulnerability to shocks and which at the same time undermine livelihoods (MoARD 2009a: 7). In this regard, the prevailing gaps in policy and the continuing decline in access to productive assets are presented as major obstacles to the growth of the agricultural sector, causing persistent food insecurity and vulnerability of farm households (Getachew 1995; Degefa 2005; Devereux 2000).

Shocks: In a discussion of shocks, Dercon (2000, 2) distinguishes between common and idiosyncratic risks by which most shocks are characterized. Common risks are "*aggregate, economy-wide and covariate risks*" that affect everybody in a particular community, whereas idiosyncratic ones affect only a particular individual in that community. He emphasizes that in practice, even in well-defined rural communities, few risks are purely idiosyncratic or purely common.

Both common risks such as climatic risks and economic fluctuations, and idiosyncratic risks increase households' vulnerability to serious hardships. However, the nature as well as the frequency and intensity of shocks people experience affect their ability to deal with the consequences. In the case of the former, Dercon (2000, 3) explains that while idiosyncratic risks can be insured within a community, such mechanisms are less likely to help in the case of common shocks. This is mainly because "*if everybody is affected, the risk cannot be shared.*" In this case, formal or informal insurance transfers (credit or insurance) from outside the community or intertemporal transfers (e.g. depletion of individual or community-level savings) might also be needed. With regard to the frequency and intensity aspect of shocks, it is emphasized that "*if bad shocks are repeated over time, then coping is more difficult*" (Dercon 2000, 3). Moreover, based on their empirical findings in Indonesia, Gertler and Gruber (2002) show that relatively small but frequent shocks such as transient illness are easier to deal with than large, infrequent negative shocks such as disability or chronic illness.

Coping strategies: Coping strategies have often been defined as a short-term and immediate response or reaction to unusual events or habitual stress or decline in access to food (Davies 1993; Birkmann 2011). The study primarily adapts the approach used in Tröger *et al.* (2011, 30) to differentiate between the term coping from adaptation and distinguishes between traditional coping mechanisms and coping strategies. Accordingly, in the case of the former, people react to certain stresses based on their experience and former learning whereas the later demands proactive decisions and new strategies to respond to unusual shocks and stresses.

In his extensive study of "*Famine and survival strategies*" in north-east Ethiopia, Dessalegn (1991) reveals the complex responses households make for economic and survival reasons, in which they often calculate beforehand the cost and benefit of their decisions to their livelihoods. In this regard, studies carried out in various contexts in different countries of the world have documented how coping strategies in the form of asset disposal are adopted at the later stages of distress, when other options, such as drawing out one's savings, changing consumption patterns and borrowing from kin groups, which have few long-term costs are no longer able to solve the crisis (Corbett 1988; Devereux 2001; Watts 1983; Dessalegn 1991).

Dessalegn (1991) argues that victims of crises depend on certain economic and survival grounds for their coping decisions. Dercon (2000) emphasizes that choosing a less risky crop portfolio has substantial consequences for the income of poor households. In his study in Tanzania he shows that households with limited liquid assets (livestock) grow proportionately more sweet potatoes, a low-return, low-risk crop (Dercon 1996). Likewise, Morduch (1990) shows that asset-poor households devote a larger share of their land to safer traditional varieties of rice and castor than to riskier but high-return varieties. The findings of Dercon and Krishnan (1996) in Tanzania and Ethiopia showed that poor farmers typically enter into activities with low entry costs: firewood collection, charcoal, collecting dung cakes, casual agricultural wage employment, etc. Entry into high-return non-crop activities, such as cattle rearing or shop keeping, is restricted to richer households with access to capital. Non-agricultural wage employment is generally restricted to those with some education.

2.2 The gender dimension of food insecurity and vulnerability in Ethiopia: state of the art

Kabeer (2003) states that gender-related constraints are usually reflected in gender inequalities in respect of access to resources and development opportunities. This has much relevance to the situation of rural women in Ethiopia. Abundant studies that focus on food security and coping strategies show the existence of a systematic link between gender and access to resources and options in livelihood strategies. The disadvantages experienced by rural women are rooted in the widely deep-seated patriarchal norms of rural institutions. Studies in different regions of the country identify factors such as the patrilocal nature of marriage arrangements, the gender division of labour, gender-biased cultural values, and widespread perceptions concerning women, which disfavour women in terms of access to resources and benefit sharing between men and women (Devereux *et al.* 2005; Aregu *et al.* 2011; Frank 1999).

In Ethiopia access to arable land is possible through the family, the state and the market (Yigremew 2005). Yet, in the absence of strong land markets and the difficulty of large-scale allocation of land through the state channel, the question regarding who gains access to family land[1], and who is marginalized, becomes increasingly important. In the study by Yared (1997), the location of residence after marriage, which is largely patrilocal, has negative implications for women's access to land through inheritance and resource division in the case of divorce. Such patrilocal arrangements also have negative impacts on widowed women who head households.

Examining patterns of parental transfers of assets on marriage in Ethiopia, Fafchamps and Quisumbing (2005) found that brides receive less land and other assets than grooms from their parents upon their first marriage. The study by Fafchamps and Quisumbing (2000), covering the Tigray, Amhara, Oromiya and South regions, concluded that women are highly vulnerable to losing their land upon divorce, and that this is related to their low asset position at the beginning of household formation. Moreover, women inherit less upon the death of their parents. With regard to the acquisition of land through informal rental land markets, the findings of Tesfaye and Adugna (2004) in the Southern Ethiopian highlands reveal that FHHs are more often on the supply side of the market as compared to their male counterparts. There is also evidence that female-headed households possess smaller plots as compared to their male

[1] Inheritance through the family takes two different forms. The first one is inter-generational transfers in the form of either inheritance after a parent's death or inter-vivos transfers, such as land gifts upon marriage or anticipated inheritance. The latter takes the form of transfer through intra-household allocation of plots to specific members (de Janvry and Sadoulet 2001).

counterparts (Dejene 1994; Etenesh 1999; Ali 2000; Kumar and Quisumbing 2012; Sara and Muluneh 2013).

Chafetz (1991) argues that women's position in the traditional division of labour, which is largely in the domestic and reproductive spheres, produces power inequalities between men and women. Such power differences further reinforce the gender-based division of labour. Chafetz also underscores that what is more damaging is the unequal ranking and unequal rewards attached to these tasks, which make women more disadvantaged in the sharing of benefits. The findings reported in Frank (1999) and Odame *et al.* (2002) show that rural women in Ethiopia must work up to twice as many hours per day as compared to men.

Aregu *et al.* (2011) emphasize the complexities in drawing general conclusions about the division of labour and the pattern of allocation of livelihood benefits among women and men in Ethiopia. Drawing on various case studies across the country, the report shows significant inter- and intra-regional variation in gender allocation of agricultural tasks, commodity marketing and profit-sharing across gender. The works of Yigremew (2000), Dejene (1994) and Bishop-Sambrook (2004) are among those that show differences in gender roles and agricultural division of labour in different regions of Ethiopia. There is enough evidence that the agricultural division of labour generally excludes women from farming since ploughing is considered a man's task. Etenesh (1999), with reference to her study carried out in central Ethiopia, points out that such assumptions are based on the cultural ideology which denies women's ability to plough the land. Studies by Frank (1999) and Yared (1997) in communities of the Amhara region, and Mogues *et al.* (2009) in the southern part of Ethiopia, in Sidama, also support these findings.

The impact of the gender-based division of labour is reflected in access to resources such as livestock possession. Livestock ownership is a very desirable wealth attribute in rural Ethiopia. In plough-based agriculture, a pair of oxen is considered a necessity. Income from livestock is also used to augment consumption and to cover production expenses, and it is a form of insurance, especially in the case of crop failure (Teklu 2003, 33). Yigremew (2001), in a study carried out in two rural communities in northwest Ethiopia, revealed that ownership of oxen is a necessary condition for access to extension programmes and to get recognition from *kebele* officials. It was also found out that female-headed households own fewer oxen than their male counterparts. The finding in Mogues *et al.* (2009), from eight districts in seven regions of the country, shows that although cattle may be jointly owned by couples, cattle ownership by women heads of households is very uncommon within the rural communities. Control over the sale of livestock and livestock products is generally gender-differentiated, with women tending to market small livestock and poultry, as well as dairy products and eggs. The sale of cattle and other large livestock is for the most part in the male domain.

In Ethiopia, the use of improved agricultural technologies is an important factor that determines agricultural production and productivity. Therefore, increasing the availability of and access to improved agricultural inputs for smallholder farmers is an appropriate strategy (CSA 2011, 9- 11). Much of the literature on agricultural technology adoption in Ethiopia has attributed the low level of technology use to a variety of social, economic, natural, and institutional factors (Bezabih 2003; Degnet *et al.* 2003; Legesse 2003; Yohannes *et al.* 1990). Empirical evidence indicates that women play an important role in decision-making in agriculture and in the adoption of agricultural technologies (Addis *et al.* 2001). Nevertheless, studies regarding the gender dimension of technology adoption show that gender roles impinge on the relationship of farmers with agricultural experts and government officials in the course of development interventions. There is evidence of the systematic marginalization of women by bureau heads, development agents and peasant associations. Various studies have also shown a gender gap in access to rural infrastructural services such as irrigation water, extension services, credit and rural institutions at household level (Mogues *et al.* 2009; Yigremew 2001; Ogato *et al.* 2009). The development interventions that target women are highly influenced by existing practices, in that extension packages tailored for women emphasize sheep and goat husbandry and poultry, for instance, rather than higher-value forms of livestock such as oxen (Mogues *et al.* 2009).

The strong link between water supply and food security and vulnerability to shocks in different regions of the country are well documented (Tesfaye and Degefa 2009; Tesfaye 2012). Yet, empirical evidence on access to water harvesting technology reveals gender based inequalities. For instance, access to water infrastructures in Alba district shows the presence of gender bias in benefiting from water harvesting technology. Level of performance in agriculture, capacity to cover costs of pond construction and own interest are identified as factors reinforcing access to the technology among FHHs, while economic and labour constraints are primarily mentioned as the prominent factors negatively affecting their access. Other factors such as lack of time, lack of knowhow, and preference for male farmers are further reasons for their lack of involvement (Rebeka 2006).

It has been well documented that agriculture and local livelihoods in Ethiopia are under serious climatic and environmental threat (Tröger *et al.* 2011). Therefore it is evident that access to off-farm activities is a critical coping strategy of the rural poor (Carswell 2000; Hussein and Nelson 1998). Rural livelihood diversification as defined in Ellis (2000, 15) is "*the process by which rural households construct an increasingly diverse portfolio of activities and assets in order to survive or improve their ability to make a living.*" They can be used as a means for coping, adaptation, and accumulation depending upon the wealth status of households. Studies on livelihood diversification strategies in Ethiopia reveal the gendered nature of these activities. A study in Southern Ethiopia, for instance shows trading as the most important non-farm activity among women, while men can resort equally to both trading and casual labour. Moreover, while non-farm activities make a high contribution to the income of poorer households which are primarily headed by women farmers, the issue of the "invisibility" of women's work as compared to men's is highlighted as a major concern (Carswell 2002, 795).

Moreover, the increasing stress on households' economy exacerbated by climatic events has led to the gradual decline of informal social networks which are the source of coping mechanisms to escape seasons of crises. As the findings of Tröger *et al.* (2011) in the Nyangatom society revealed, such institutional transformation presents certain challenges to women who depend more on the reciprocal relationships to cope with food shortage seasons. Women in Nyangatom society also face other institutional barriers due to their gender roles in the course of carrying out activities such as exchanging animals like sheep or goats against sorghum or other food items. They are generally described as more vulnerable to food security as compared to their male counter parts.

2.3 A historical consideration of rural development approaches and gender studies

Research on women or gender-and-development issues requires a thorough understanding of both development and feminist theoretical frameworks, as both development and feminist scholarship have contributed to shaping thinking and policy (Connelly *et al.* 2000). In the section that follows, an attempt is made to touch on some of the rural development discourses as they have gradually evolved under the umbrella of the dominant development paradigms[2], starting from the decades following the post-war era till the present time. In a parallel manner, an attempt is made to capture the major contributions in the field of gender studies within the evolving rural development paradigms.

Rural development paradigms and gender studies (1950-1980)

Prior to the 1960s, rural development thinking seldom focused on smallholder agriculture. Rather, mainstream economic development thinking for this period focused on industrialization, fuelled by models of surplus labour (Ashley and Maxwell 2001). However, in the 1960s, development thinking redefined the importance of small-scale agriculture as the ultimate engine of growth and development. This thinking said that in low-income countries, small-scale agriculture plays a key economic role by providing labour, capital, food, foreign exchange and a market for consumer goods for the emerging industrial sector. There is a large body of literature starting in the 1960s and going on well into the 1990s, arguing in favour of small-scale agriculture and its immense importance in terms of bringing economic development (Schultz 1964; Berry and Cline 1979; Lipton and Longhurst 1989). During this period, the efficiency of small farms in exploiting new green-revolution technologies through large-scale state investment in infrastructure, research, and support for the adoption of new technologies was found highly convincing (Ashley and Maxwell 2001). In general, it is worth noting that despite the shifts in rural development thinking in the 1950s and 1960s, in the eyes of rural development experts the rural poor were nothing but passive recipients of development benefits. In other words, many of the development schemes and interventions were to a large extent guided by a top-down approach.

From a gender perspective, these projects have double-impacted rural women, since this group made the bottom of the development recipients' list. In other words, much of the rural development planning during this period marginalized rural women in the South, thus contributing to a wider poverty gap between men and women. Development interventions were meant to serve mainly men who were seen as potential candidates for turning Western development projects into economic growth in the South. During this period, development in general, and rural development schemes in particular, ignored rural women, who were generally assumed to play the role of housewives and whose (re)productive role in the economy is categorized as merely informal (Connelly *et al.* 2000).

In the decades that followed, there was increasing recognition of the failure of the top-down approach to meet the anticipated development objectives. Thus, the 1980s and 1990s witnessed the second paradigm shift in rural development discourses, characterized by a major shift from the top-down or "blueprint" approach towards the bottom-up, grassroots or "process approach." These decades also saw swift changes in the approaches towards understanding rural development by giving due emphasis to a participatory process that targeted rural people as the centre of every development intervention (Chambers and Conway 1992, 3).

Among some of the key features of such changes are the advent of farming systems research (FSR), growing acknowledgement of the validity of Indigenous Technical Knowledge (ITK), the rise of the participatory method, such as rapid rural appraisal (RRA) techniques in the

[2] The decades following the post-war era have witnessed a number of emerging rural development paradigms within the larger development themes, such as modernization, state intervention and market liberalization, as well as participation and empowerment in that chronological order. For a detailed reading of the evolution of these discourses, see Ellis and Biggs (2001).

1980s, and participatory rural appraisal (PRA) and participatory learning and action (PLA) during the 1990s. On top of that, an "actor-oriented" perspective gave the poor a distinct status, thus transforming them from passive recipients to active participants in development. Later on, large-scale involvement of the government in the agricultural sector was considered a barrier, due to the structural adjustment and market liberalization programmes that were launched from the beginning of the 1980s. At the same time, the weakening role of governments in various social and economic arenas led to the rise of the era of NGOs as agents of rural development.

Although achievements vary across regions, the period under discussion left a lasting legacy in the form of the rise of gender as an area of concern in rural development (Ellis and Biggs 2001). Gender issues were subjected to close scrutiny. This led to the differentiation of men's and women's livelihood strategies based on the gender-defined agricultural division of labour, which is also manifested in domestic roles, off-farm opportunities and responsibility for protecting the welfare of family members at times of adverse external events (Ellis 2000).

This era was known for the new feminist movement. Sparked by Boserup's (1970) seminal work on "women and economic development"[3], women's issues took the centre stage worldwide with the launch of what is known in history as "The United Nations Decade for Women" in 1975 at the conference held in Mexico City with the theme "Equality, Development and Peace". A goal was established to integrate women into development processes in the decade ahead. That included enabling women to have a say in policy design and implementation (Moser 1993, 141). From this period onwards, remarkable developments can be observed in relation to the emergence of new paradigms and analytical frameworks guiding research and policy on gender issues. A major advance in this regard was the emergence of the theoretical approaches known as "Women in Development" (WID) in the 1970s and 1980s, and "Gender and Development" (GAD) since the 1980s, the two most important approaches to integrating women in development (Connelly *et al.* 2000, 57-61).

Under its leading motto, "equal opportunity for women", the WID approach emphasized the importance of including women in the development process. The WID approach advocated enhancing economic development through integrating women into welfare-oriented development projects and plans. However, by embracing Western institutions as the sole means to improve the life of women in the South, the WID approach largely ignored women's indigenous knowledge. Preoccupied with notions of modernization and efficiency, the WID approach has done little to address important factors in women's subordination, such as race and class (Connelly *et al.* 2000, 57).

Rural development paradigms and gender studies in the sustainable development era

Rural development discourses in the last decades of the 20th century and in the years following the onset of the 20th century were preoccupied with a growing interest in livelihood and sustainability concepts. These studies found their intellectual inspiration in understanding the lives of poor people and thus set out to adopt an actor-oriented perspective in development studies (Chambers and Conway 1992). This view was able to influence rural development policies. The sustainable livelihoods framework was introduced for the first time by the Brundtland Commission on Environment and Development as a unique approach to rural development in the early 1990s (Krantz 2001). In the subsequent years the Department for International Development (DFID) incorporated the sustainable livelihood approach in its international development cooperation programmes. Indeed, it increasingly attracted global atten-

[3] Her work uncovered the largely ignored but instrumental economic roles played by rural women in the agricultural sector in the South.

tion to become an important developmental paradigm (Lele 1991; Goudie 1998; Chambers and Conway 1992).

The era also witnessed a major policy shift in the form of the so called "structural adjustment programmes" with its "gender blind" economic models. Economic policies followed under the structural adjustment program increased women's workload, perpetuated the traditional gender division of labour, and maintained the notion that women are naturally suited to caring work (Elson 1995, 1863). These changes in socio-economic realities in various countries around the world led to the emergence of a new group of feminist scholars with a new perspective called the GAD approach.[4] This framework is also known as the Empowerment Approach.

According to the GAD framework, women are not mere recipients of development, but active agents who are able to pursue their own interests. In this regard, a distinction is drawn between women's interests[5] and gender interests.[6] The latter are further categorized as practical or strategic (Molyneux 1985).[7] Another defining peculiarity of the GAD framework lies in its focus on the relationship between women and men, rather than on women *per se*. In general, in the 1980s, donor agencies and state machineries consolidated their WID activities, but the GAD perspective increasingly shaped the interests and activities of feminist NGOs and was in turn shaped by their experiences (Connelly *et al.* 2000).

This theorizing on women and particularly the GAD perspective is of great relevance to the current widely used theories in the field of development studies, including the sustainable livelihoods framework. In the author's opinion, some of the notions that are emphasized in the GAD framework, such as decision-making by women, gender interests, and gender power relations, relate to the actor-oriented and demand-driven approach of the sustainable livelihoods framework (SLF), which is one of the analytical tools adopted in this thesis.

The increasing popularity of the GAD approach has raised doubts regarding the analytical depth of the SLF in respect of explaining social issues such as gender inequality, and has contributed to shaping further thinking. For instance, several gender-based livelihood studies have revealed gender-based biases in areas of health care, education, food consumption, access to property and assets and division of labour.[8] These studies showed that access to livelihood resources as emphasized in the SLF is not gender-neutral. Therefore, current trends are towards creating a synergy with other fields within the social sciences. A good example in this case is De Haan and Zoomers (2005). The authors have attempted to deepen the SLF by integrating notions of power from other disciplines and fields of gender studies.[9]

In general, it can be asserted from the discussion above that the dominant development paradigms and theories on gender and development that have shaped research policy and action in

[4] This approach emerged in the 1980s as an alternative to the WID and Women and Development (WAD) approaches.

[5] A biological category that assumes homogeneity.

[6] A socially constructed set of relations and material practices.

[7] Practical gender needs arise out of concrete conditions; these are immediately perceived needs, such as the need to provide food, shelter, education, and health care. Strategic gender interests arise out of an analysis of women's subordination and require changes in the structures of gender, class, and race that define women's position in any given culture. Strategic interests include the goal of gender equality.

[8] On healthcare and food, see, among others, Harriss 1990; Drèze and Sen 1995, and Agarwal 1986; on education, see Drèze and Sen 1995; on property, see Agarwal 1994; and on gender division of labour, see Dasgupta and Maiti 1986.

[9] For instance De Haan and Zoomers (2005) have pointed out that power is an explanatory variable to the understanding of social relations, institutions and organizations that determine livelihood opportunities for the poor. Thus, they propose that the SLF can draw a lot of insight from the field of gender studies.

the South, generally echoed Western culture, knowledge and ideology. Thus, they reflect common concerns of "the politics of identity" as described in Connelly *et al.* (2000, 104). That being said, without overlooking all the positive outcomes that have been achieved so far, theorizing in gender and development has faced some serious odds. In other words, despite the various analytical frameworks adopted and the significant achievements in terms of production of research, knowledge, policy and action in the course of the gender and development debate so far, its goal of coming to grips with crucial issues such as gender inequality is still far from practical realization. For instance, the WID approach, trapped by the efficiency argument of modernization theory, has largely focused on participation to meet women's practical needs and has ignored women's strategic interests that are equally essential in order to achieve women's equality with men. The empowerment approach of GAD, even though it holds huge promises towards the goal of achieving gender equality, has been hampered by the continuous manipulation of development agencies.[10]

2.4 Sustainable livelihoods approach (SLF)

A number of theories have been used to explain food shortages at various geographical scales which range from global to individual. The widely cited theories include the Food Availability Decline (FAD) (Millman and Kates 1990; Devereux 1993), the "Food Entitlement Decline" (FED) (Sen 1981), the "Political Economy Explanations" (Devereux 1993), a theory that considers food shortages as a disaster (Wisner *et al.* 1994), and the "Sustainable Livelihoods Framework" (SLF) which looks at food insecurity as an outcome of vulnerable livelihoods (Davies 1996; Carney 1998; Scoones 1998; Ellis 2000). Generally, the existing literature shows that there have been shifts in ideas on food security from food scarcity to livelihood vulnerability perspectives. Considering its contextual applicability, this thesis adopts the SLF as one of its analytical and organizational tools.

Chambers and Conway are known for their widely used definition of livelihood: "*a livelihood comprises capabilities, assets (stores, resources, claims and access) and activities* (Chambers and Conway 1992, 6)." Furthermore, sustainability within the sustainable livelihoods framework is understood as "*a function of how assets and capabilities are utilized, maintained and enhanced so as to preserve livelihoods*" (Chambers and Conway 1992, 10). Thus, in this definition, sustainability embraces concerns beyond ecological conservation to include social elements that make up people's livelihoods. Following their analysis, sustainability raises two fundamental questions; that of environmental sustainability and social sustainability. The former gives emphasis on the "external" impacts of livelihoods while the latter deals with the "internal" capacity to respond to outside pressure (Chambers and Conway 1992, 6-10). In relation to the above explanations of livelihoods and sustainability one finds a slightly more advanced version in Carney (1998, 2): "*A livelihood comprises the capabilities, assets (including both material and social resources) and activities required for a means of living. A livelihood is sustainable when it can cope with and recover from stresses and shocks, maintain and enhance its capabilities and assets, both now and in the future while not undermining the natural resource base.*"

The summary in Farrington *et al.* (1999) sheds light on the underlying premises and the methodological interpretation of the SLF, based on its primary assumptions. According to them, the sustainable livelihoods framework is based on the idea that people enjoy a range of liveli-

---

[10] Further, evidence of such analytical failures embedded in their conflicting conceptualization and operationalization was found by Whitehead & Lockwood (1999, 525-555) in a review of six World Bank poverty assessment reports for Uganda, Ghana, Tanzania, and Zambia. They found that without a clear understanding of gender, poverty assessment was driven by epistemological and methodological choices for measuring poverty, as well as a set of prescriptions for reducing poverty.

hood outcomes (health, income, reduced vulnerability, etc.) by drawing on a range of assets to pursue a variety of activities. The activities they adopt and the way they reinvest in asset-building are driven in part by their own preferences and priorities. However, they are also influenced by the types of vulnerability they experience, including shocks (such as drought), overall trends (for instance, resource stocks) and seasonal variations. Options are also determined by the structures (such as the role of governments or the private sector) and processes (such as institutional, policy and cultural factors) which people face. In aggregate, these conditions determine their access to assets and livelihood opportunities and the way in which these can be converted into outcomes. In this way, poverty, and the opportunities to escape from it, depend on all of the above (Farrington *et al.* 1999, 1). Figure 2.1 presents the sustainable livelihood framework.

Figure 2.1 Sustainable livelihoods framework

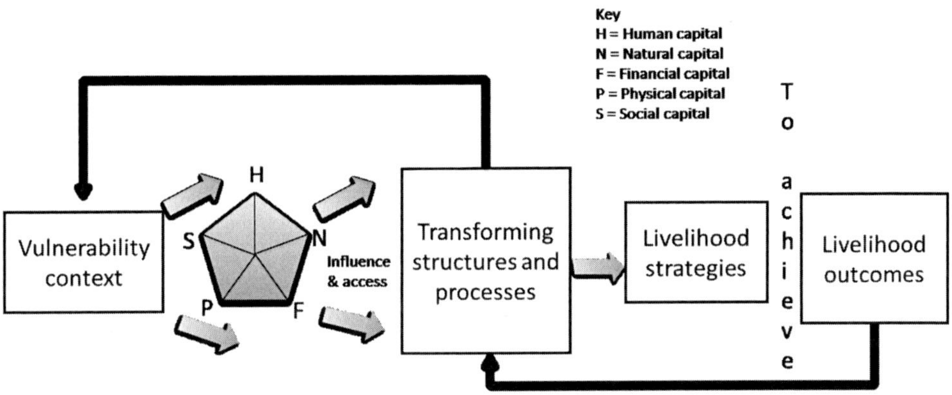

Source: DFID (Department for International Development), 1999

In this thesis, the SLF is primarily used to organize analysis of the complex livelihood system of households in Meskan district. However, it is essential to note that this study adopts the framework with due recognition of its insufficiency with regard to addressing the gender power relations which are crucial to this study. Thus, the subsequent section draws on the works of certain contemporary critics as well as the conceptual contributions made so far to strengthening its analytical depth in this respect. Then, some of the conceptual and methodological meanings attached to notions such as access to livelihood assets, livelihood strategies and livelihood outcomes, which are emphasized in the framework, are dealt with in detail.

Critics and recent conceptual contributions to SLF

The sustainable livelihoods framework is a response to the demand for a holistic approach to understanding livelihoods and poverty (ODI 1998; Ashley and Carney 1999; Chambers and Conway 1992 and others). The approach is widely recognized as a way of coming to grips with the complexity of livelihoods and opening possible entry points for development interventions. The purpose of this approach, as described by Appendini (2001, 24) cited in de Haan & Zoomers (2005, 30) is "*to search for more effective methods to support people and communities in ways that are more meaningful to their daily lives and needs, as opposed to ready-made, interventionist instruments*".

However, there is a large body of literature that provides a critical assessment of the applicability of the sustainable livelihoods framework for researchers and development practitioners

(Ashley and Carney 1999; Krantz 2001; Scoones 1998; Lele 1991 and others). Scoones, one of the most influential contributors as well as critic of the sustainable livelihoods approach, claims that while the approach apparently overweighs the value attached to the role of institutional and organizational settings in understanding the livelihoods of the poor, it has failed to spotlight the crucial and yet unfortunately missed notions of power and power relationships that are embedded in the transforming structures and processes (Scoones, 1998). Similar criticisms can be found in the work of Ashley and Carney (1999, 35), where they write: "*The SL framework overall can convey a somewhat cleansed, neutral approach to power issues. This contrasts starkly with the fundamental role that power imbalances play in causing poverty.*" In the light of the above observation, Ashley and Carney (1999) emphasize poor people's access to strategic assets while de Haan and Zoomers (2005) specifically use the notion of trade-offs in commanding resources that the sustainable livelihoods framework envisages, to emphasize how the SLF undermines the notion of power relations.

Another criticism is found in Ellis (2000). He praises the livelihoods approach for providing a clear account of the non-separable linkage between assets and the choices that people have in pursuing their livelihood strategies. However, he criticizes the fact that such definitions fail to show the dynamism in access to resources and opportunities that result from changing norms and events in the social and institutional contexts which play a critical role in this respect.

Taking a similar stand, a very notable and contemporary contribution to the SLF is found in de Haan and Zoomers' (2005, 38) work, "*Exploring the Frontier of Livelihood Research.*" Their contribution, which is the product of years of application of the framework in livelihood studies, builds on earlier studies of institutions and processes. It also goes a step further to deal specifically with the notion of "power" that is integral to the "transforming structures and processes." In their article, livelihood is understood as multi-goal driven, diversified in context, and multi-located. Thus, they emphasize that to be a powerful analytical tool in livelihood studies, the SLF needs to broaden its scope in addressing questions of access and decision-making. They strongly promote the view that by borrowing ideas from gender and development studies, as well as from the field of sociology, the sustainable livelihoods framework can strengthen its conceptual basis as well as fill its gaps. They argue that the framework needs to go through a process, which they identify as "*getting to grips with the concept of power relations.*" Understanding power relations is thus considered a key to unlocking the institutions and processes. This leads to revealing the real factors that guide social relations, which are the medium through which access to resources is determined.

This thesis draws certain methodological and theoretical lessons from de Haan's and Zoomer's contemporary analysis of the sustainable livelihoods framework. The study benefits from their contribution for the reason that their approach gives a particular emphasis to power relations embedded in social relations in the understanding of people's livelihoods. In a similar manner, in this thesis, the concept of gender power relations is considered essential, since such relations are an integral part of the daily livelihood routines of individual men and women and lead to livelihood outcomes that can be categorized as either positive or negative.

From the discussion above, it is clear that there is an ongoing scientific inquiry to further deepen the conceptual and analytical foundations of the livelihoods framework. Therefore, the researcher finds it important to further highlight some of these contributions for an advanced methodological application of the sustainable livelihoods framework. Accordingly, the subsequent section elaborates some of the concepts contained in the SLF, such as assets and access to livelihood assets, livelihood strategies and livelihood outcomes, by drawing on some of the recent theoretical and conceptual contributions made in livelihood studies.

Assets and access to assets

Assets, sometimes also referred to as livelihood resources or capital, are the building blocks or the basis for various activities from which livelihoods are constructed. In general, the SLF divides assets into five main types. These are natural capital, physical capital, human capital financial capital and social capital. Natural capital means the natural resource base (land, water and air) from which services are derived in the production of livelihoods. Physical capital refers to assets brought into existence by economic production processes. Such assets may include tools, machines, irrigation canals, etc. Human capital refers to the skills, knowledge, ability to labour, good health and physical capability essential to the successful pursuit of different livelihood strategies. Financial capital refers to the capital base (cash, credit, savings, and other economic assets such as basic infrastructure, production equipment and technologies) which are essential for the pursuit of a livelihood strategy. Social capital means the social resources (networks, social claims, social relations, affiliations, associations) upon which people draw when pursuing different livelihood strategies that require coordinated action (Scoones 1998).

Assets can also be categorized as tangible and intangible. The former consist of concrete assets such as land and livestock, whereas the latter are more abstract, including claims and access, relations that can be relied on for material, moral or other forms of support at times of stress or shock (Ellis 2000; Scoones 1998; Chambers and Conway 1992). Claims may be made on individuals or agencies, on relatives, neighbours, patrons, chiefs, social groups or communities, or on NGOs and the international community, in the framework of programmes for drought relief or poverty alleviation. They are based on combinations of rights, social conventions, moral obligation and power. Access means the opportunity in practice to use a resource or service, or obtain information, material, technology, employment, food or income (Chambers and Conway 1992, 7-8).

From the above definition it follows that the notion of assets captures not only the ability of individuals to fulfil their material needs; it also refers to the individual's state of well-being, the set of interactions that actors engage in, as well as the kind of power involved in such relationships. In the words of Bebbington (1999, 2022):

"A person's assets, such as land, are not merely a means with which he or she makes a living: they also give meaning to that person's world. Assets are not simply resources that people use in building livelihoods: they are assets that give them the capability to be and to act. Assets should not be understood only as things that allow survival, adaptation and poverty alleviation: they are also the basis of agents' power to act and to produce, challenge or change the rules that govern and control the use and transformation of resources."

In the sustainable livelihoods framework, assets interact with institutions and processes in which access to such assets is directly influenced by these institutions and processes. However, the framework seems to oversimplify the process of this interaction which in reality presents a very complex picture. The complexity surrounding the notion of access within the sustainable livelihoods framework has captured the interest of various researchers. For instance, de Haan and Zoomers (2005, 34) state "*access depends on the performance of social relations and these are sometimes far from harmonious*." Drawing on the work of Leach *et al.* (1999)[11], they give a more advanced explanation of access using the notions of endowments and entitlements[12] , mapping[13] and capabilities. Accordingly, actors, to transform a specific

[11] Leach *et al.* (1999) work primarily draws on Sen's Entitlement theory.

[12] Entitlement simply refers to the right individuals enjoy, whereas endowment is what one actually owns.

asset into an endowment and eventually into capabilities, have to go through a process of endowment mapping. Important in this process are the sets of utilities that people enjoy by wielding a legitimate command over goods or services: their entitlements. Such entitlements eventually translate in to capabilities, depending on various economic and social variables. For instance, actors can enjoy certain entitlements through their village membership status, household membership or other roles in their communities (Leach *et al.* 1999, 233 cited in de Haan and Zoomers 2005, 34-35). Thus, "*access is not only an issue affecting the use or acquisition of capitals: it is also an issue associated with the beneficial exploitation of livelihood opportunities.*" (de Haan and Zoomers 2005, 34).

Livelihood strategies

The sustainable livelihoods approach is a very helpful tool to identify the different livelihood strategies households employ based on the resources they are able to access. The SLF envisages three broad clusters of livelihood strategies which are seen as options open to rural people. These are: agricultural intensification/ extensification, livelihood diversification, and migration. Rural households gain more from agriculture (including livestock rearing, aquaculture, forestry, etc.) through processes of intensification (more output per unit area through capital investment or increases in labour inputs) or extensification (more land under cultivation), or they diversify to a range of off-farm income-earning activities, or move away and seek a livelihood, either temporarily or permanently, elsewhere. More commonly, they use a combination of these strategies, together or in sequence. Such a combination can be referred to as a 'livelihood portfolio' (Scoones 1998; Serrat 2010).

In Chambers' and Carney's 1992 analysis of the determining factors for livelihood strategy choices in households, it is highlighted that livelihood strategies can be "predetermined by accident of birth." For instance, traditional gender roles clearly demonstrate the role of gender in determining livelihood activities. De Haan and Zoomers (2005, 35) also strongly emphasize the gender feature of livelihood activities. They consider livelihood activities as "engendering processes of inclusion and exclusion."

Within the SLF the focus on household heads as active decision-makers rather than passive victims, and as those whose collective active involvement responds to and enforces change, has led to the assumption that livelihood activities are a result of the decision making of households whose members share common goals. However, against the backdrop of significant social, economic and ecological changes[14] that have been observed in the livelihoods of people over the years, de Haan and Zoomers (2005, 38-40) challenge the above assumption by introducing what they call strategic and unintentional decision-making in livelihoods. This can be better captured through the concepts of styles[15] and pathways[16], two important notions used to "disentangle regularities" as shown in their discussion. In general, their discussion on the decision-making behaviour of people in accessing livelihood opportunities has the poten-

[13] 'Mapping' refers to the process by which endowments and entitlements are shaped, which makes the meaning of mapping much closer to access.

[14] Heterogeneous household units, diversification of livelihoods increasing mobility of household members are mentioned as reasons for increasing divergence of individual goals and household goals, challenging the credibility of the label household strategy (see de Haan and Zoomers 2005, 38).

[15] Styles are "*distinguishable patterns of orientation and action concerning the variety of means to achieve security; these patterns are structured by an internal logic and conditioned by the social, economic and personal characteristics of the people involved*"(Nooteboom 2003, 54 cited in de Haan and Zoomers 2005, 40-41).

[16] A pathway can be defined as "*a pattern of livelihood activities which emerges from a co-ordination process among actors, arising from individual strategic behaviour embedded both in a historical repertoire and in social differentiation, including power relations and institutional processes, both of which play a role in subsequent decision making*" (de Haan & Zoomers 2005, 45).

tial to broaden the understanding of livelihood strategies, going beyond the perception of strictly planned activities to show the role of social relations, institutions and organizations in shaping the decision-making of actors over access to livelihood resources. In this way the interaction between institutions and assets and livelihood strategies, which is presented as a simple and direct relationship in the SLF, can be grasped much more closely.

Livelihood outcomes

According to the SLF, livelihood strategies lead to livelihood outcomes that are sustainable. Various organizations that apply the SLF use their own measurement of development outcomes. For instance, the DFID Sustainable Livelihood Guidance Sheet has the sustainable use of natural resources, income, well-being, vulnerability, and food security. On the other hand, Oxfam uses income, well-being, vulnerability, food security, sustainable environmental resources, and security of the non-use value of nature, while CARE (Cooperative for Assistance and Relief Everywhere) uses security of food, nutrition, health, water, shelter and education, community participation and personal safety. In general, livelihood outcomes such as food security and well-being are mentioned by most agencies.

In the process of operationalizing the SLF, the greatest challenge is how to achieve these outcomes without risking greater trade-offs. In this regard, it is essential to point out two types of argument as summarized in Chambers and Conway (1992). These consist of those which show concern from the point of view of social sustainability and those which prioritize environmental sustainability. The former advocate the importance of the increased capacity of livelihoods to withstand outside pressure in a form of enhanced ability to cope with stress and shocks, thus ensuring livelihood sustainability. The latter raise issues related to the external impact of livelihoods on other livelihoods (Chambers and Conway 1992).

2.5 Understanding livelihood through the sustainable livelihood framework and the "theory of structuration"

This section presents a brief discussion of the "theory of structuration". The first subsection discusses some key concepts within the theory in order to give a rough picture of it. Following this, the subsequent section presents the basic concepts of power and power relations within the structuration theory. This is important for two reasons. First of all, understanding livelihoods through gender power relations is vital to illuminating the gender gap in households' access to livelihood assets and their vulnerability. Secondly, adding the notion of gender power relations fills the gap in the SLF.

2.5.1 The "theory of structuration"

The struggle people make to meet their needs in constructing their livelihoods leads us to the question: are the poor, who are emphasized in the SLF, free to act as they please, or are they shaped and governed by outside structures? This is a question that has always been challenging to political and social scientists interested in studying society. While some scholars have found the answer in human agency, arguing in favour of a 'naturalistic' standpoint, others have drawn insights from the field of interpretative science, underscoring the role of structures in explaining human behaviour.

Against the above backdrop, this thesis rejects the two polarized positions in favour of a theory of action where social systems can be understood in terms of the mutual power relationship between structure and human agency, namely "the theory of structuration"[17]. The process of

[17] Anthony Giddens' "theory of structuration" (1979) is among the most influential philosophical works of the late 20th century in social science. According to Giddens, his work is a response to the lack of a theory of action

structuration is the field of mediation between agency[18] and structure. The founder of the structuration theory, Anthony Giddens, advances the view that there is mutual interaction between human agents and social structures instead of these being two conflicting entities operating independently. During this mutual interaction, human agents produce, reproduce or modify social structures via their actions and social practices in the course of their everyday life. Giddens emphasizes that the social roles and practices of actors are themselves the realization of broader institutions (Giddens 1984, 25). Similarly, as stated in Berger and Luckmann (1967, 176), the existence of institutions can be determined by the agents on the ground which continually acts to bring them to life. At the same time, the presence of institutions continues to shape the practices of actors.

Actors generally find themselves involved in social spaces consisting of sets of rules, norms, codes, obligations and meanings that need to be interpreted constantly. During the process of "structuration" they prefer one norm or social order over the other, depending on their own understanding and perception and judgment of their everyday life situation. Such orientations are guided by what Tröger (2002, 172) after Giddens (1997, 78) calls "perception of everyday life." For instance, during stable periods that mark the continuity of social life, actors typically orient themselves by taking up widely used and locally popular norms and values. As a result, they rely on what Giddens calls their "practical consciousness" (Giddens 1979, 24). This is typical of actors' everyday routines. However, such a level of reflection is unlikely to rescue them at times of structural disintegration, disturbance or contradiction. This is when actors find themselves confronted with a range of possible norms to choose and work with, and they are able to do so only if they act depending on their "discursive consciousness". However, not every actor is capable of such articulation at a discursive level. An actor's degree of understanding of the conditions in his environment may depend on his or her mental ability to comprehend the existing information. The result of such interaction at a discursive level is the modification, substitution or production of new structural elements which influence future actions and social practices (Giddens 1984, 49).

In Giddens' terminology, such interaction forms "a duality of structure" where the structure consists of both the rules and resources (Giddens 1984, 374). The duality of structure represents the essential reclusiveness of social life, tied up in social practices that stretch across space and time (Giddens 1979; 1984). Here, the space/time notion is given an emphasis since it is an essential and inherent feature of the constitution of social practices: as agents move and interact they meet at stations or definite time-space locations within bounded regions (eg. Home, street, cities) (Giddens 1984, 112). These regions what Giddens prefers to call, "locale", define the setting of interaction (Giddens 1979, 207; 1984, 118). The nature of these interactions will have implication to people's capabilities (Giddens 1984, 112).

Constituted with the reproduction of social life are the rules and norms that permanently enable, constrain and give meaning to social practices. These rules and norms embedded in social practice are referred to as institutions (Scott 2008; Giddens 1984). A comprehensive discussion of institutions is found in Scott (2008), where an institution is presented as consisting of regulative, normative and cultural cognitive elements. As a regulative entity, institutions act as codified rules, decision-making procedures, and monitoring and sanctioning mechanisms that together guide human conduct. As normative systems they consist of norms and values. Norms specify how things should be done and define the legitimate means to pursue valued

---

in the social sciences that discusses institutional power and agency, time and space relations that are inherent in the constitution of social interaction (Giddens 1979, 2-3). As such, it is an approach to social theory that attempts to reconcile the theoretical dichotomies of agency/structure, subjective/objective, and micro/macro perspectives that feature in most of the dominant schools of thought in social science (Giddens 1979; 1984; 1993).

[18] Understood, according to Giddens, as the ability of an individual "to act otherwise" (Giddens 1984, 14).

ends. Norms also decide upon actors' social roles, their obligations and privileges (Scott 2008, 52-55). Values and norms together help to define the objectives of actors and the most appropriate ways to achieve them. As cultural cognitive entities, institutions work at the deepest level of society. These consist of socially constructed symbolic representations such as words, signs or gestures. Actors using these symbolic representations, break down reality into comprehensible categories that they eventually translate into what they assume is the appropriate social practice. These are collectively shared in a social group and are simply taken for granted. They shape the meaning that actors assign to objects, practices and places, as well as to their own identities. Capturing this cultural-cognitive dimension of institutions requires the inclusion of emotional aspects such as confusion and disorientation, or confidence and certitude, in institutional analysis (Scott 2008, 59).

2.5.2 Power and power relations in social practices

With regard to the structuration theory, in order to know the power of agents it is necessary to understand the resources available to them. Resources, in Giddens' interpretation, are "the media whereby transformative capacity such as power is employed in the routine course of social action; but they are at the same time structural elements of social systems as systems, reconstituted through their utilization in social interaction" (Giddens 1984, 92).

All social positions, within social systems, are also "power positions" in the sense that they are integrated within reproduced relations of autonomy and dependence (Giddens 1979, 119). These power relations or relations of autonomy and dependence are reciprocal: no matter how asymmetrical the distribution of resources involved, all power relations involve a relation of autonomy and dependence "in both directions" (Giddens 1979, 149). Thus, according to the structuration theory, even the most dependent actor or party in a relationship retains some autonomy (Giddens 1979, 93). No matter how subordinate an actor may be in a social relationship, the very fact of her/his involvement in that relationship gives her or him a certain amount of power over the other (Giddens 1979, 6). This is inevitable unless that person has lost all her/his capability of action due to some sort of confinement and through supervision. But in all other cases, in other words in all cases in which human agency is exercised within a relationship of any kind, power relations are two-way (Giddens 1979, 149).

According to the structuration theory, the incorporation of power as integral to the constitution of practices is an important element in the conceptualization of human action. Action involves interventions producing certain outcomes and such action logically involves power in the sense of transformative capacity to achieve those outcomes (Giddens 1979; 1984; 1993). Power is generated by definite forms of domination, in a parallel way to the involvement of rules with social practices, and can be identified as an integral element or aspect of those practices (Giddens 1979, 69). Therefore it is understandable that the agent deploys a range of casual powers in the flow of daily life and is also able to influence those deployed by others. Action depends upon the capability of the individual to 'make a difference' to a preexisting state of affairs or course of events. An agent ceases to be such if he or she loses the capability to make a difference, that is, to exercise some sort of power (Giddens 1984, 14).

2.6 Understanding livelihood vulnerability using the SLF and the theory of structuration: a gender perspective

Vulnerability is a versatile term and has been defined and used in various contexts (refer to section 2.1). The vulnerability concept within the sustainable livelihoods framework points to the external environment in which people live. For example, trends in population growth, national and international economics, natural resources, politics, and technology; sudden shocks or events such as health problems, earthquakes, floods, droughts, conflict, agricultural problems such as pests and disease, economic shocks; and the seasonal vulnerability of prices, production, employment opportunities can all have an impact on livelihoods (Chambers & Conway 1992, 11). Mostly, these events are beyond people's direct and indirect control and can have a negative effect on access and control over assets. It is also obvious that the external dimensions of vulnerability such as shocks, stresses and crises are common phenomena affecting both men and women.

The SLF also emphasizes actors as agents and active decision-makers who constantly respond to external events. However, the agency exercised by actors is not free from external influences. For instance, it depends on the availability of means (access, assets) as well as the ability to mobilize assets and employ various coping strategies without heavy loss. This implies that the capacity to withstand hardships can vary from individual to individual, making the degree of vulnerability of one individual or group more challenging than others. This argument can be further strengthened by Kabeer's (2003) assertion in which she claims that even though class, poverty, ethnicity and physical location may interact to challenge the state of equality between individuals, the gender factor tends to make these constraints more severe, leading to greater inequality. Hence, the internal dimension of vulnerability, which this thesis primarily deals with, can be reinforced or constrained depending on the nature of power relations in any social interaction that involves both men and women. This brings forth the need to examine structures and institutions that constitute gender power relations, for which ideas have been borrowed in this thesis from the theory of structuration. Strengthening the above arguments, Kelly & Adger (2000) assert that an in-depth understanding and examination of structures and institutional constraints should be an essential process in the analysis of vulnerability.

The lives of women and men tend to be shaped and reshaped through their gendered decisions and livelihood practices. Following Giddens (1984, 374) "duality of structure", it can be asserted that such gendered actions continuously produce and reproduce gendered structures that constitute power at different levels. Various norms, rules, societal orders and social practices ensure the continuation of unequal gender power relations as long as there is no radical proactive decision to disrupt the process. Gender disparities supported by gendered social practices generally have a negative impact on the livelihoods of women, leading to their unique experience of vulnerability in terms of food security as well as their capacity to cope with stresses and shocks.

There are various mechanisms by which such unequal gender power relations embedded in various patriarchal rural institutions manifest themselves. Agarwal (1997) explains that gender power relations embedded in traditional and cultural norms, as well as customary practices, put restrictions on women's access to resources and their decision-making ability beyond the household level. In other words it is meant that both men and women are involved in claims, bargaining and negotiations both inside and outside the domestic sphere, in choosing livelihood options, benefiting from resource appropriation, accumulation, transfer, distribution, and so on. In the meantime, the effects of unequal gender power relations works against the interests of women in various dimensions. For instance, biased perceptions constrain women household heads from taking up a wide range of livelihood activities both inside and outside the domestic sphere if, by tradition, these tasks are assumed to be in the men's domain. Women usually find themselves in weak bargaining positions when it comes to claiming their rights over their resources, due to traditions that look down on women who try to confront men. As a result, at household level female-headed households are more sensitive to livelihood shocks and stresses. Moreover, they lack the capacity to employ various strategies to cope with such perturbations, and have a greater tendency to suffer from the consequences of such crises. These, as a result, undermine their household food security, leaving them with a fragile livelihood.

In this thesis, the SLF and the structuration theory are combined to facilitate the conceptualization of vulnerability of households which leads to an understanding of the causal factors of food insecurity and ability to cope with livelihood shocks. To facilitate such a conceptualization of vulnerability, the dynamic interaction, social processes and relationships between local actors and the existing rural institutions is given primary attention. Gender power relations are implicit in many of the existing structures, as well as social processes and interactions. The degree of power involved during social interactions and relationships determines the capacity of one individual to take advantage of the resources at their disposal. This leads to a positive or negative livelihood outcome, for instance, a strong or weak capability to ensure one's food security, as well as the ability or inability to effectively cope with adversities. In such a way the vulnerability of households in the study area is captured at the required level. The analytical framework below summarizes the discussion so far.

2.7 Conceptual framework for data analysis

The analytical framework presented in Figure 2.2 below identifies the multiple attributes related to vulnerability, food security and coping strategies and illustrates the directions or pathways of their interconnections. The key terms used in the conceptual diagram are vulnerability, institution, access and entitlement to resources, agency/decision making, and coping strategy and food (in)security. Gender power relations are assumed to be an integral part of each and every component of the conceptual framework. It is assumed that vulnerability is shaped by institutions in the face of ongoing environmental changes and social and economic transformations. In the conceptual framework, it is envisaged that there are two-way relationships between institutions and actors' agency or decision-making. Actors' knowledge and interpretation of the institutions determines their orientation in exercising their agency. In the face of competing formal and informal institutions, socially embedded informal norms and customary practices dictate entitlement and access to resources, and access to resources in turn reinforces norms and customary practices. Access to resources influences actors' agency during social encounters and interactions in the social environment. In reciprocity, actors' agency defines access to resources which eventually translates into food security. In the analytical framework, people's ability to cope with shocks is the product of the interaction between people's agency, institutions and access to resources under a given temporal and spatial frame.

Figure 2.2 Conceptual framework to examine the interaction between vulnerability, food (in)security and coping strategies

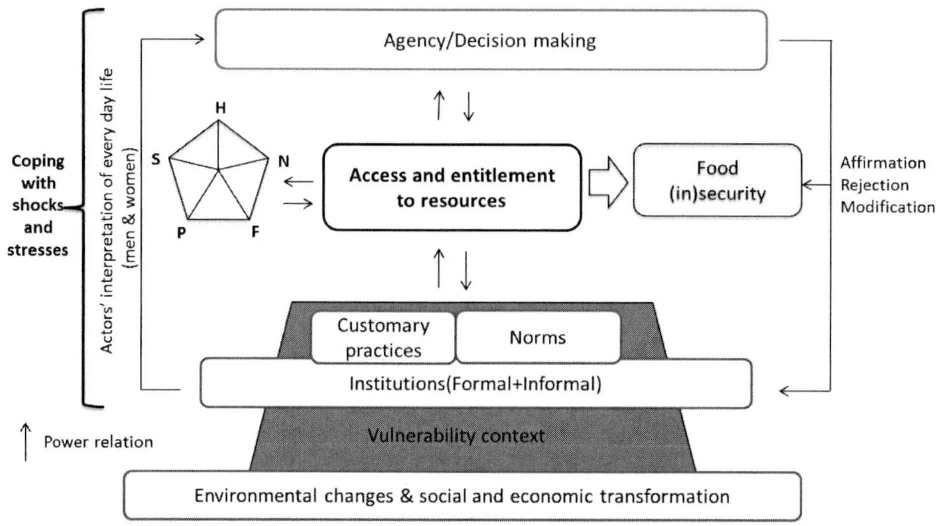

H= Human resources, S= Social resources, P= Physical resources, F=Financial resources, N= Natural resources

Source: Own design

The following section briefly discusses the main categories of analysis considered in the study in light of the conceptual framework shown in Figure 2.2.

Explaining food insecurity and coping strategies in the face of vulnerability: applying the conceptual framework lens

As shown in Figure 2.2, in the face of simultaneously existing, competing and at times conflicting governing structures such as formal and informal institutions, deep-seated patriarchal norms and customary practices are assumed to primarily shape the vulnerability context of women farmers in general and female-heads of households in particular. In this study, informal institutions, such as traditional marriage and family, gender roles in division of labour, and customary judiciary systems, are considered the major institutions that define, regulate and legitimize entitlements and claims to resources upon which decision-making and the actions of men and women depend (see chapter 7 for a detailed discussion).

Based on the categories of assets indicated in the conceptual framework, the study considers land, oxen, labour, time, water and agricultural technologies such as credit, fertilizer and improved seeds as the major livelihood assets in its empirical investigation (see chapter 8 for detail reading). It is also assumed that individuals with a specific gender can enjoy a greater degree of entitlement over these assets than the others simply by virtue of their positioning within the existing institutions. In other words, where gender-based perceptions and practices are typical features of societal life, women's rights may be determined by the interaction between different norms and customary laws that coexist with statutory law. In this regard, even in cases where there is no formal discrimination, women's rights may be restricted in practice; factors such as cultural perceptions of women's role in the family and in society, and/or female seclusion practices, can constrain women's entitlement to assets and their agency.

In Figure 2.2, both food security and coping strategies against shocks are essential livelihood outcomes at the household level. The attainment of food security and effective response at household level is not simply a function of equal access to resources and opportunities; it is also a matter of being able to exercise adequate agency to benefit from those rights, resources and opportunities. And this is facilitated by open and responsive institutions, since they could promote better mobility and flexibility, and enhance the capacity of men and women to adequately respond to vulnerability (Adger *et al.* 2002). Actors exercise their agency as they involve themselves in acts of bargaining, negotiation, deception and manipulation, subversion and resistance, either at an individual or a collective level (Kabeer 1999). In the rural context, men and women farmers continuously engage in decision-making, both inside and outside the domestic sphere, and they do so from either a weak or a strong bargaining position or power position. For instance, they exercise agency when they engage in various labour-sharing and land transaction arrangements, when they participate in community organizations, public resource use, various on-farm and off-farm activities and when negotiating over benefits from the sale and purchase of agricultural or non-agricultural commodities. In this regard, the study considers various traditional coping mechanisms and coping strategies employed by farm households (these aspects are discussed in chapter 9).

Within the context of the conceptual map shown above, institutions shape and at the same time are shaped by thinking and perceptions which continuously produce and are reproduced by the agency of men and women. During this process, individual men and women actors refer to the societal norms most applicable to their own life situation. At times, they also act by rejecting or modifying these institutional structures. Such interaction dictates the mechanisms by which gender-based stereotypical views and attitudes are reproduced and the way vulnerability contexts are shaped.

3 Appraisal of the national food security and gender policies in Ethiopia

Food as a human right got international formal recognition, even though with no binding obligations, back in 1948 through the establishment of the Universal Declaration of Human Rights (UDHR) (FAO 2009). As of May 2012, the International Covenant on Economic, Social and Cultural Rights (ICESCR) with its 156 State members, including Ethiopia, took a significant positive step by making the right to food a rule of binding international law for the States parties. Article 11 deepened the concept of the right to food by mentioning "...the right to adequate food", adding "...the fundamental right of everyone to be free from hunger..." Accordingly, the article made it clear that the state would commit itself to ensuring that its people do not starve, at the very least[1] (UN 2012).

With or without the recognition of food as a human rights issue, poverty eradication has been the leading agenda of the government of Ethiopia over the last two decades. Therefore, all of the country's development policies and strategies are geared to eradicating poverty and dependence on food aid in a relatively short period of time (MoFED 2010). The country has thus embarked on major national development plans, such as the Sustainable Development and Poverty Reduction Programme (SDPRP), which was implemented from 2002/03 to 2004/05, a Plan for Accelerated and Sustained Development to End Poverty (PASDEP), that ran from 2005/06 to 2009/10, and currently the Growth and Transformation Plan (GTP) for the period 2010/11-2014/15. Overall, great strides, including macro-economic reforms directly or indirectly aimed at alleviating the problem of food insecurity in the country, have been made.

In this chapter the focus is on two major policy areas in two separate sections. The first section presents a review of the national food security strategy as one of the most important initiatives in the history of rural development programmes in Ethiopia. Particular attention is paid to the Productive Safety Net Programme (PSNP) and the Household Asset Building Programme (HABP) as the two major strands of the national food security strategy of the government that are relevant to this study. The second section is devoted to national policies on gender. Here, an attempt is made to highlight the major legislative measures taken by the government towards improving the life of rural women, followed by an appraisal of the National Policy on Ethiopian Women (NPEW). In the case of the latter, particular emphasis is laid on the first and second National Action Plan documents on gender equality in Ethiopia. Finally, reflections are made on the relationship between gender policies and the national food security programme in respect of their content, adequacy and gaps in implementation. In other words, an attempt is made to show the extent to which gender-specific vulnerabilities are considered in the national food security programme.

3.1 Policy on food security

The first initiative directly targeted at tackling the problem of food insecurity in Ethiopia dates back to the year 1996, in which the first national food security strategy paper was issued. The strategy was further improved and updated by the government in 2002 to address the multi-dimensions of food security, by targeting chronically food insecure, moisture deficit and pastoral areas. However, the common objective of both food security papers is to attain food security at the household level, with Agricultural Development-Led Industrialization (ADLI) as the leading policy framework (Workneh 2008).

The food security strategy paper has four major components which are interlinked: the Direct Food Production Intervention Programme, the Voluntary Resettlement Programme, the In-

[1] www.ohchr.org/EN/ProfessionalInterest/Pages/CESCR.aspx accessed on 2/12/2013

come Diversification Programme, and the Productive Safety Net Programme (Workneh 2008). The new Food Security Programme (2010-2014) incorporates two new elements: the Households Asset Building Programme (HABP) and the Complementary Community Investment programme (MoARD 2009a). The food security strategy is also framed by supply (availability)[2] and demand (access)[3] dimensions in three broad environmental settings: adequate moisture environment, moisture deficit environment and pastoral environment (Workneh 2008). This paper will focus on the second setting which the study area falls under.[4]

Among all these aspects of the food security strategy, the following section selectively discusses the PSNP and the HABP, as both of them have been operationalized in the study area.

3.1.1 The Productive Safety Net Programme (PSNP)

The PSNP has been implemented since 2005 at the national level. The main reason for the Ethiopian government to initiate the programme is the desire to address the basic food needs of food insecure households via a social safety net scheme, rather than through a system dominated by emergency humanitarian aid. It is considered an effective tool for achieving food security objectives in a relatively short period of time (MoARD 2006).

The primary objective of the Productive Safety Net Programme is provision of transfers to the population in chronically food insecure districts in order to prevent asset depletion at household level. The programme is also believed to support a rural transformation process by preventing long-term consequences of short-term consumption shortages; encouraging households to engage in production and investment activities; enhancing household purchasing power; and promoting market development. The programme consists of two major components: public works and direct support. The former is implemented by mobilizing able-bodied rural households to engage in labour-intensive community-based activities. Payments to participants in public works are made both in cash and in kind, depending on the accessibility of local markets. Even though most of the public work activities are carried out on community lands and using community resources, certain activities can also be carried out on private lands. For instance, less abled households or households who have shortage of labour such as female-headed households are also assisted. In the case of direct support, an unconditional transfer of cash or food equivalent to the public work payments is provided where its beneficiaries are vulnerable households with no able-bodied members, destitute households with no means of income and the disabled or orphans *(ibid)*.

More than seven million people have received PSNP transfers since its inception in 2005, enabling them to meet consumption needs, reducing the risks they face and providing them with alternative options to selling productive assets (MoARD 2009a, 8). The PSNP initially targeted 262 districts in eight regions of the country. The SNNPR, where the study district is located, is one of the beneficiary regions. The selection criteria for beneficiary households include: households that have faced food shortages for at least three consecutive years and those that have been continuously receiving assistance, households that are vulnerable due to extreme asset loss, households with no alternative sources of income through relatives and family members or membership in a community organization (MoARD 2006; 2009b).[5]

---

[2] Refers to the availability of food through increased domestic production.

[3] Refers to ensuring access to food through increasing the purchasing power of food deficit households.

[4] See 4.3 for more information on rainfall distribution in Meskan district.

[5] Districts currently defined as chronically food insecure are those targeted by the PSNP up until 2009. Their eligibility for the PSNP in 2005 was defined by the frequency with which they required food assistance in the ten years preceding the design of the PSNP (the ten years up to 2004) (MoARD 2009b).

3.1.2 The Household Asset Building Programme (HABP)

As one part of the government's food security strategy, the Household Asset Building Programme is specifically aimed at increasing the source of income and asset holdings of food insecure households. It is designed to improve the food security status of male and female members of chronically and transitorily food insecure households who have a food gap of three months or more in either a normal or moderately bad year. The HABP programme provides the same services to households in the PSNP programme and to those who have graduated from the PSNP and are not yet food secure. However, in the case of capacity and/or resource constraints the first priority of the HABP will be those within the PSNP and those who have recently graduated from the PSNP. Even though both female- and male-headed households are equal beneficiaries, due consideration is given to women and female-headed households who have specific needs in relation to confidence building and skills such as financial literacy (MoARD 2009c).

In the course of its implementation, supportive policies were implemented in a parallel manner. Two of the most relevant for this study include the rural finance policy and the women- and youth-related policies. In the case of the former, a strategy was designed to improve the poor financial practices seen in the previous implementation period, through the introduction of financial services [credit provided by the programme through Micro finance institutions (MFIs) and Rural Saving and Credit Cooperatives (RUSACCOs)]. According to the national food security policy document, providing adequate, appropriate and timely credit is considered one of the means to break the poverty cycle and bring about sustainable development (MoARD 2009c).

Rural cooperatives were assumed to play a key role in this regard. Accordingly, women are encouraged to form cooperatives[6] of their choice and benefit from saving and credit schemes in order to increase their income from on-farm and off-farm activities. This is also meant to overcome the problems women face in accessing credit through formal MFI and private money lenders. In order to ensure that women and the youth benefit from the development programmes, "developmental packages" were designed.[7] The packages aim at boosting the benefits obtained by women and youth from their labour on the land. They also have the goal of ensuring that agricultural development programmes and extension efforts, off-farm activity promotion programmes and market development programmes help women and youth to organize themselves in saving and credit cooperatives (MoARD 2009c, 9).

The HABP as the new component of the government's food security programme for the period 2010-2014 is designed to tackle a variety of factors related to poor households and food insecurity in a more effective manner than the previous food security components. The major areas where improvements are expected, according to the policy document, are the functioning of rural financial services, better linkage between programmes and institutions through the provision of complementary services such as technical follow-up, inputs and animal health services and increasing physical access to effectively functioning markets as well as market information (*ibid*).

A number of government bodies ranging from the highest federal level to the lowest *kebele* administration level are involved in the planning, implementing and monitoring of both the PSNP and the HABP (see Table 3.1). Workneh (2008) has identified certain weaknesses re-

[6] Cooperative promotion in Ethiopia is governed by Proclamation No. 147/98, which defines a cooperative as "a society established by individuals on voluntary basis to collectively solve their economic and social problems and democratically manage same" (FDRE 1998).

[7] The packages were developed by the Ministry of Women Affairs and the Ministry of Youth and Sport.

garding the structural setup presented in the PSNP document. These include the overlapping activities, vagueness of the policy document on the division of tasks and decision-making roles, the inclusion of structures that are not functioning in the PSNP and lack of clarity whether there is any operational linkage among the various bodies.

Table 3.1 Organizational structures implementing the Productive Safety Net Programme (PSNP) and Household Asset Building Programme (HABP)

Administration level	Implementing Offices	Programmes run by the office
District	District Development Committee (WDC)	HABP
	District Food Security Task Force (WFSTF)	HABP, PSNP
	District Food Security Desk (WFSD)	HABP
	District Agriculture and Rural Development Office(WOARD)	HABP, PSNP
	District Office of Finance and Economic Development (WOFED)	HABP, PSNP
	District Cooperative Promotion Office	HABP
	Inputs and Marketing unit	PSNP
	All district Sector Offices	HABP, PSNP
	Early Warning and Response Desk	PSNP
	Natural Resource Desk	HABP
	NGOs	HABP, PSNP
Kebele	*Kebele* Council and Cabinet	HABP, PSNP
	Kebele Development Committee (KDC)	HABP
	Kebele Food Security Task Force (KFSTF)	HABP, PSNP
	Kebele Appeals Committees (KACs)	HABP, PSNP
	Community Food Security Task Force (CFSTF)	HABP, PSNP
	Rural Savings and Credit Cooperatives	HABP
	Development Agents (DAs)	HABP; PSNP
	Clients and non-client households	HABP

Source: Compiled from PSNP and HABP policy documents (MoARD 2009b; 2009c)

3.2 Government measures towards gender equality and women empowerment

Agarwal (1994) describes empowerment as "*a process that enhances the ability of disadvantaged (powerless) individuals or groups to challenge and change (in their favour) the existing power relationships that place them in a subordinate economic, social and political position*" (Agarwal 1994, 39). Kabeer underpins this definition by describing women's empowerment as "a process by which those who have been denied the ability to make strategic life choices acquire such ability" (Kabeer 1999, 437). The fulfilment of such conditions is assumed to assist the disadvantaged to participate in, negotiate with, influence, control, and hold accountable institutions that affect their lives.

So as to improve the life of Ethiopian women in general and rural women in particular, several legislative and policy measures have been pursued by the Ethiopian government in the last two decades. Ethiopia is party to most international human rights instruments[8] including the Committee on the Elimination of Discrimination against Women (CEDAW), under which the country is committed to ensuring the elimination of all forms of discrimination against women. This includes ensuring that women enjoy equal civil, cultural, economic, political and social rights just like their male counterparts. The Federal Democratic Republic of Ethiopia (FDRE) has proved its compliance with the CEDAW by taking ground-breaking measures in both the legislative and policy areas to ensure women's equality with men. The following section discusses both legislative and policy aspects, followed by an appraisal of these measures in terms of their gaps and their contribution to alleviating the vulnerability and food security of rural women.

3.2.1 Legislative measures

Article 35 (1) of the 1994 FDRE constitution guarantees women equal rights with men in the enjoyment and protections of rights provided in the constitution. It also prohibits laws, customs and practices that oppress or cause bodily or mental harm to women. The constitution makes provision in article 9, sub-paragraph 4 that the rights provided in the constitution are in conformity with the international instruments. The same article states that the constitution is the supreme law of the land and cannot be contravened by any law, customary practice or decision of an organ of state or public official (FDRE 1995).

In addition to these provisions, so as to keep national legislation in line with international standards, the constitution proclaimed a revised family law in the year 2000. This law has replaced the discriminatory provisions of the 1960 civil code. In this regard, article 34 of the constitution provides that men and women have equal rights while entering into, during and on termination of marriage. This included equal division of all assets between husband and wife upon divorce (FDRE 2000).[9] The revised family law, which is in accordance with these constitutions, recognizes marriage through civil, religious and customary channels. The provisions include the same minimum legal age for men and women, the right to marry with free and full consent of the intending spouses, prohibition of bigamy or polygamy[10], joint management of the family[11], protection of women in relation to the proceedings and effects of divorce[12], and the rights of women in irregular unions[13]. Moreover, the revised criminal law explicitly outlawed all forms of violence against women, including rape, abduction, female genital mutilation (FGM), early marriage and other harmful traditional practices (FDRE 2005b). The revised family law and criminal law are effective both at federal and regional levels.

Under article 35 (7), the constitution also guarantees women equal rights with men with respect to the use, transfer, administration and control of land and explicitly states women's equal rights with regard to access to land (FDRE 1995). Accordingly, the 2005 rural land administration proclamation confirms the equal rights of women in respect of the use, administration and control of land, as well as in respect of transferring and bequeathing holding rights (Part two, Art. 5 (4). A land administration law enacted by each regional council shall

[8] The International Convention on Civil and Political Rights (ICCPR), International Covenant on Economic, Social and Cultural Rights (ICESCR) and others

[9] Between 2000 and 2005, only Amhara, Oromiya and Tigray implemented the law, but as of 2011 all regions have been made to apply the revised family law (Hallward-Driemeier and Gajigo 2013, 6).

[10] Article 11 of the revised family law

[11] Article 50 of the revised family law

[12] Articles 80, 81, 82 and 83 of the revised family law

[13] Articles 102 and 103 of the revised family law

(Part Two, Art. 6) ensure free assignment of holding rights…, without differentiation of sexes… (Sub-Art. 1), and lay down a system based on transparency, fairness and the participation of peasants, especially of women, for purposes of assigning holding rights and carrying out distribution of holdings (Sub-Art. 10) (FDRE 2005a).

The right to equal treatment during inheritance is reaffirmed under article 35 (7) of the constitution by emphasizing the equal enjoyment of rights in practice by men and women. The right of women to vote and be elected and women's equal access to full participation in power structures and decision-making is also affirmed under article 3 of the constitution. Consistent with the new constitution and the CEDAW, articles 41 and 35 of the constitution proclaim equal access to education with temporary special measures to be adopted to compensate for past policy mistakes (FDRE 1995). In the sphere of employment, Labour Proclamation No. 377/2003, Articles 87 and 88 outlawed discrimination on grounds of sex in employment and payment (FDRE 2004).

3.2.2 The National Policy on Ethiopian Women (NPEW)

According to the country's gender profile of 2002, the National Policy on Women (1993) was one of the initial ground-breaking steps taken by the transitional government of Ethiopia, showing its commitment to the advancement of women in the country (Haregewoin and Emebet 2002). The policy takes several factors as its major premises, such as the poor human rights records of Ethiopian women, the lack of recognition given to their contribution in sustaining family and society, their vulnerability to poverty and natural and man-made disasters, deprivation of equal opportunities and gender biases in the society. Thus the policy firmly declares that development processes that exclude women are destined to partial success.

The main objectives of the National Policy on Ethiopian Women include creating and facilitating conditions of equality between men and women, creating conditions to make rural women beneficiaries of social services like education and health, and eliminating stereotypes, prejudices, customs, discriminatory perceptions and practices that are based on male supremacy and enable women to take part in decision-making process at all levels. Strategies have been designed to achieve the above objectives. The policy identified eight major areas that need to be dealt with, including detailed strategies to accomplish the above objectives. The eight broad areas of the policy are:

1. Ensure the full development and advancement of women and guarantee them the enjoyment of their human rights.
2. Modifying or abolishing discriminatory laws, practices and customs and ensuring the equal participation of women in economic and social and political life as well as in decision making processes at all levels.
3. Creating awareness to change discriminatory attitudes towards women.
4. Ensuring the participation of women in the formulation and implementation of laws, policies, projects and programmes that concern and benefit them.
5. Encouraging research on ways of lightening the workload of women and increasing their income.
6. Coordinating and following up the implementation of all government programmes concerning women to avoid duplication.
7. Incorporating women's affairs into the government structure and establishing offices in all ministries and government organs.
8. Ensure that all government laws and policies are not discriminatory, provide protection for the rights of women, give special attention to the conditions of rural women

and make sure that women participate in and enjoy the fruits of development activities on an equal basis with men.

For its implementation, three bodies were identified; The Women's Affairs Office (WAO) in the Prime Minister's Office, the regional Women's Affairs Sectors and the Women's Affairs Departments in ministries and public organizations, each implementing organ having its own duties and responsibilities. The WAO at national level, being accountable to the Prime Minister's Office, has the task of coordinating, facilitating and monitoring women's affairs activities at the national level. It is the highest body entrusted with the responsibility of issuing laws and policies towards attaining policy objectives. The regional Women's Affaire Sector (WAS), which is accountable to the regional administrative council, facilitates and coordinates activities related to women affairs at the regional level. In addition to this, the office is responsible for making sure that gender concerns are included in the preparation of plans, researches and projects in the region. Depending on the situation, the office can engage in initiating and formulating policies. The Women's Affairs Department (WAD) is accountable to the ministry or public organization under which it is established with the same powers as the other departments. It is the organ responsible for implementing the national women's policy (WAO and WB 1998).

With the above brief review of the objectives and structures of the NPEW, the following subsection attempts to assess the policy and its implementation by placing special emphasis on its role in improving the life of rural women, because that is the area that is relevant to this thesis. This will be discussed within the context of the country's major poverty-reduction programmes.

In line with the priority areas set out in the policy, the responsible organs have undertaken different activities, such as movements to put an end to discriminatory and harmful practices against women, efforts to educate the public through the mass media, attempts to ensure the inclusion of gender concerns in policies and programmes, forums to exchanges views and information among governmental organizations, NGOs and other stake holders, and trainings to build the capacity of the implementing organs.[14] Given the fact that the NPEW was prepared ten years earlier than the establishment of the first national poverty-reduction programme[15], it did not directly address the issue of poverty up until the time of its action plans in (2002-2006) and (2006-2010). During this period, however, the action plan explicitly declared poverty reduction as one of the critical areas to promote gender equality (Sosena and Tsahai 2008, 101).

During the implementation period of the first National Action Plan on Gender Equity (NAP-GE) (2002-2006), important measures were taken to tackle gender inequality and improve the livelihood of rural women. The economic empowerment of women was emphasized to enable women to secure independent and sustainable livelihoods, and ensure that they can fairly participate in the social and economic development of the nation. These include measures towards improving rural women's food security and preventing further vulnerability through increasing women's access to agricultural land, employment opportunities, agricultural inputs, extension services, improved food crop production and animal husbandry, as well as lightening the work-load of women (*ibid*).

[14] Unpublished reports by Progynist 2002
[15] This refers to the "Sustainable Development and Poverty Reduction programme" implemented from 2002/03 to2004/2005.

The second action plan of the NPEW (2006-2010) coincides with the implementation period of the government's second development programme called the Plan for Accelerated and Sustained Development to End Poverty or PASDEP (2005/6-2009/10). With continuing political will and commitment, this plan placed strong emphasis on the participation of women in political, economic and social affairs, and the organizational structures responsible for women's affairs were strengthened at federal, regional and local levels. The family law and criminal law were amended to fully ensure the rights of women. The NAP-GE focused on seven priority areas that the PASDEP documents identified as major areas to be tackled in order to ensure gender equitable development. The most significant of these were the participation and benefits of women in agricultural packages, ensuring that equal benefits from land rights accrue to husbands and wives, and the issuing by regional governments of land utilization certificates bearing the names of both husband and wife. Moreover, in food security programmes vulnerable children and female-headed households were given priorities (MoFED 2010). Unlike the first action plan, the government tried to integrate the second plan in PASDEP and made all ministries responsible for implementing women's five-year action plans including allocation of budget (MoFED 2006). In general the content of the NAP-GE (2006-10) was improved over the previous action plan, such as the provision of refined indicators for effective implementation, inclusion of environmental issues, and integrating the plan into PASDEP (Sosena and Tsahai 2008, 101).

The government of Ethiopia in its current development programme paper, the Growth and Transformation Plan (GTP) envisages sustaining and improving the achievements that were registered during the previous PASDEP period in the areas of gender and poverty reduction. For instance, the programme document envisions providing special emphasis to women and youth by increasing the proportion of women and youth beneficiaries of the agricultural extension to 40 percent (MoFED 2010, 47). Increasing the number of women entrepreneurs that graduate from operating micro enterprises to small and medium scale businesses, increasing the number of women who receive training in management and entrepreneurship, and increasing the number of women beneficiaries of credit and saving services are also at the heart of the development strategy. In addition, it envisages the abolition of harmful traditional practices that adversely affect the active engagement of women in decision-making and that have a negative effect on the welfare of women in general (MoFED 2010, 110).

For the current GTP period, three implementation strategies are identified with regard to tackling the issue of gender inequality. The first stresses strengthening women's associations and organizations. Secondly, a more conducive environment will be created for women to actively participate and benefit from development programmes through their associations and organizations. Finally, for the participation of and benefits to women to be effective and sustainable, effective coordination of women's associations and organizations and all other actors are assumed to be critical.

3.3 Assessment of the implementation of rural development policy and gender policies

Following the above brief review of legislative and policy measures within the context of rural development in Ethiopia, it can be said that the Women in Development (WID) and Gender and Development (GAD[16]) movements which were born in the 1970s and 1980s have contributed a lot to framing development policies on women in the last two decades. For in-

[16] GAD recognizes women's reproductive role, while maintaining that the traditional division of labour is by and large socially constructed and can therefore be changed, and that the exclusion of women in development is due not to any natural laws but to the man-made division of labour, socio-cultural attitudes and ideologies of oppression.

stance, shifts in thinking from women's projects (the WID approach) and the integration of women in all areas of economic, social and political life through the GAD approach or the empowerment approach have impacted development policies and programmes. Accordingly, policies and programmes were geared towards the call for greater involvement of women and men in the development process by integrating women's practical and strategic needs (Sosena 2000).

Within such a framework, some of the approaches followed in the food security strategy, and in the various legislations and policies on women formulated by the FDRE, can be seen as a breakthrough towards empowering women in general and rural women in particular. However, the aim here is to pinpoint some of the major shortcomings in the existing provisions for reducing the vulnerability of women to food insecurity and enhancing their livelihoods.

Workneh (2008) states that even though various government structures and committees have been established to coordinate and implement PSNP and other food security programmes, experience shows that there is a serious limitation in terms of the roles and responsibilities of the individuals who are assigned to various tasks. As a result, appeals, complaints and dissatisfaction, combined with lengthy processes to get just decisions, severely constrain households with a low social position and weak bargaining power in their community, such as female-headed households.

Two major policy gaps can be identified in relation to policies concerning women. The first is related to the failure of the organizational structure to get down to the grassroots level. And the second is related to the relationship between the different policy implementation organs. In the case of the former, the fact that the organizational system of implementation of women's policy stops at district level and fails to go down to *kebele* (the smallest administrative unit in Ethiopia) level, has constrained its effectiveness in reaching its main beneficiaries, rural women. In the case of the latter, the lack of a direct relationship between women's affairs departments at district level and the food security strategy implementing body at *kebele* level constrains the effective implementation of the food security strategy. In this regard, the lack of a clear mandate for the district Women's Affairs Department to monitor gender mainstreaming[17] including the incorporation of gender equality and women's interests within the PSNP is a major missing element. This leads to the implementers' lack of the necessary gender knowledge that needs to be taken into consideration in the implementation of programmes.

Moreover, regardless of the provision of various legislative reforms discussed in 3.2.1, attaining gender equality both at macro and micro levels remains a formidable task in the country. One of the major institutional gaps identified as a constraint to the empowerment of women is the institutional plurality with regard to law enforcement at various administrative levels. In this respect, gender experts and women's rights activists criticize the constitution from two major standpoints. The first one is related to the recognition given in the constitution to informal institutions, such as customary and religious laws defining, regulating and adjudicating

[17] "Mainstreaming gender perspective is the process of assessing the implications for women and men of any planned action, including legislation, policies or programmes, in all areas and at all levels. It is a strategy for making women's as well as men's concerns and experiences an integral dimension of the design, implementation, monitoring and evaluation of policies and programmes in all political, economic and societal spheres so that women and men benefit equally and inequality is not perpetuated. The ultimate goal is to achieve gender equality" (UN ECOSOC 1997). http://www.un.org/documents/ga/docs/52/plenary/a52-3.htm accessed on 13/12/2013.

personal status. These institutions have a significant influence in the areas of marriage, divorce, property ownership and division, child custody, inheritance and adoption. The second one relates to the autonomy given to the regional states, with power to enact family codes, while all other matters of civil law remain within the jurisdiction of the Federal State. As a result of these factors, in a highly traditional and patriarchal society like Ethiopia, the challenge imposed by such institutional plurality is generally reflected in the poor enforcement of laws and legislations. This is visible particularly in the lower and micro-level administration units such as *districts* and *kebeles*.

In addition to the above stated limitations, the enforcement of existing anti-discrimination policies and laws is often under-resourced, especially at the sub-national level. Assessments made over the years indicate that the implementation organs are not performing effectively due to lack of human and material resources (WAO and WB 1998).

Generally, the efforts to ensure the elimination of discriminatory perceptions and practices against women through policies and legislation are a good move towards equal enjoyment of human rights by women, which is also an instrumental tool to attain food security for both men and women farmers. Yet, changing entrenched patriarchal social practices and values often requires a continuous pro-active approach with high-level political commitments and monitoring mechanisms. Thus, based on the lessons learned in the implementation period of the women's policy, efforts should continue to revise the policy with due attention to its proper implementation. The task of protecting rural women from further poverty and vulnerability continues to face formidable challenges.

4 Description of the Study Area

4.1 Geographic setting

Meskan district is situated in the northern part of the Gurage zone of the SNNPR. It is one of the 15 districts of the zone. The district has an area of 50,177 ha (roughly 502 sq. km) and its elevation ranges from 1501-3500 m above sea level. It is situated between 7.99-8.27°N latitude and 38.26-38.57°E longitude. The landscape consists of 10 percent mountains, 55 percent plateau and 35 percent rugged topography. There are two broad traditionally recognized agro-climatic zones, namely the *dega* (cold zone) comprising 20 percent of the district and *woina dega* (mild temperate) covering 80 percent (Meskan District Finance and Economic Development Office 2011).

The district town, Butajira, is 133 km south of the capital city, Addis Ababa. The District Administrative Council, sector ministries and the Gurage Zone Development Association have their offices in Butajira. The district town is connected to Addis Ababa by an all-weather gravel road. The district has 40 *kebeles*, the lowest administrative units. The *kebeles* where this case study was carried out, namely, Beresa, Dobo Tuto, Dirama and Semen Shershera, are located relatively close to Butajira. The four *kebeles* are in the same agro-climatic zone that is woina *dega* (mild temperate). Topographically, they are characterized by rugged, mountainous and rocky landscapes. The location of Meskan district and the study sites within the various administrative regions of Ethiopia is shown in Figure 4.1.

Figure 4.1 Location of Meskan district and the study sites

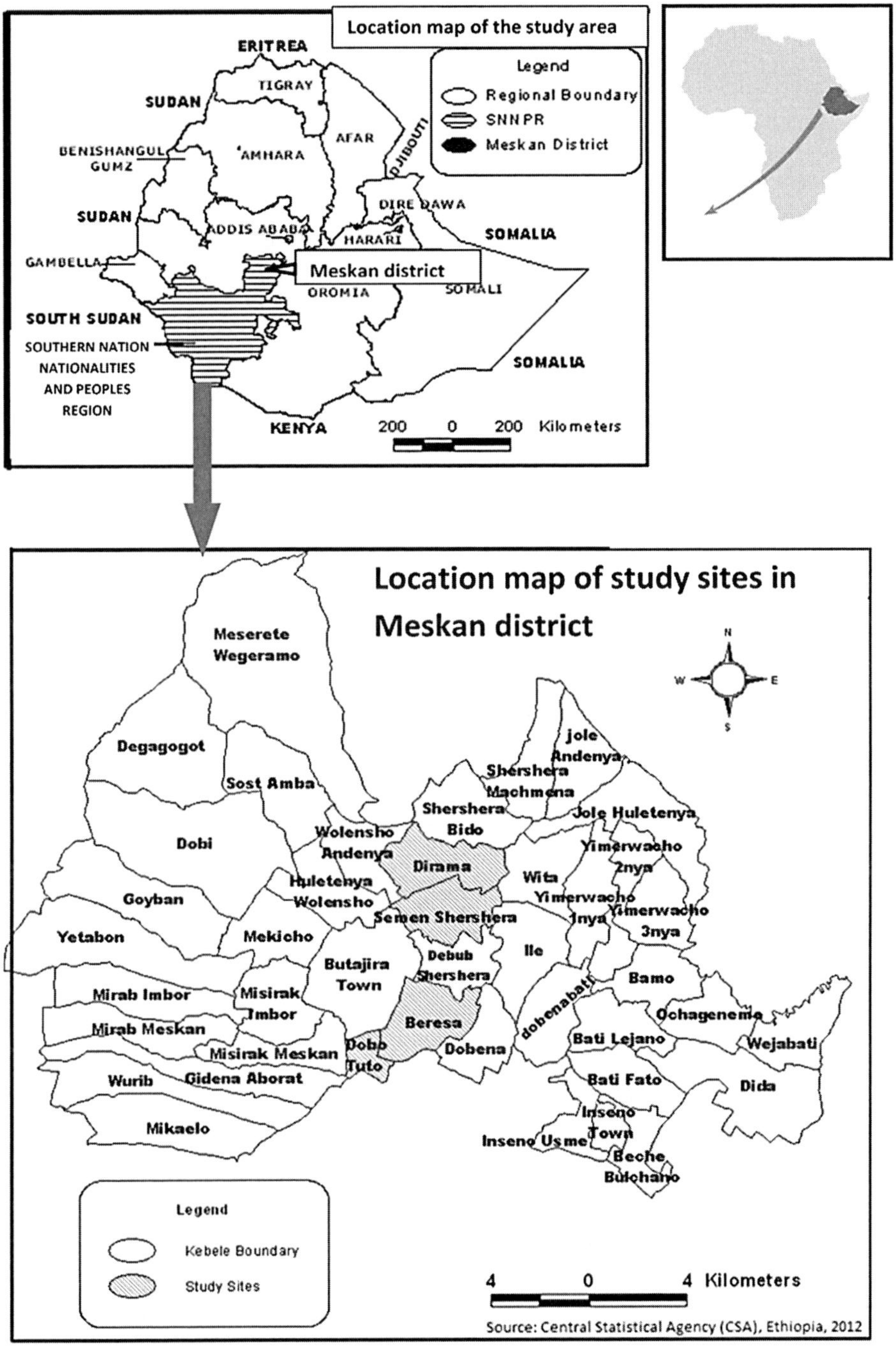

4.2 Demographic and socio-economic background

According to the data obtained from Meskan district information bureau in the year 2011, the total population of the district is estimated to be around 174,647, of which 92 percent are engaged in agriculture. Out of the 35,211 households in the district, 22,913 are male-headed households while 12,298 are female-headed households. The average household size is 5 persons per household. The fertility rate for women of productive age is one of the highest in the country. The population is predominantly Muslim, comprising nearly 61 percent of the district population. Orthodox Christianity follows with 34.2 percent and Protestantism (4.7 percent). Polygamy is common among the Muslim population. The dominant ethnic group in the district is Gurage, comprising 82.6 percent of the population (Meskan District Finance and Economic Development Office 2011). The Gurage speak a Semitic language composed of clusters of dialects. Linguistically, they are divided into three groups: the northern Gurage (Aymellel or Sodo), the eastern Gurage (Silte Wellene, Zeway, etc.), and the western Gurage (mainly Sebat Bet, i.e. "seven houses" or tribes, and Gogot (Meskan) (Worku 1996, 133-134). Rural families live in *tukuls* (traditional round houses made of wood and plastered with clay, covered by thatched roofs). The majority of rural families share their living quarters with their domestic animals (Figure 4.2).

Figure 4.2 Traditional *tukuls* of the Gurage

Source: Field work, 2011

The study sites possess one of the least developed physical, social and economic infrastructures in the country. Even though there have been improvements in recent years, in terms of expanding rural health centres, communication, or electric power infrastructures, the coverage in the rural areas is still very low. For instance, the information obtained from the District Finance and Economic Development Information Dissemination Office shows that there are only one hospital, two doctors and 52 nurses in the district serving the entire population. Hydroelectric power has reached only 24 out of the 40 *kebeles* in the district, and only 8 of them have direct access. The coverage of potable water in rural areas stands at 42 percent (Meskan District Finance and Economic Development 2011). Based on the observations made in the study *kebeles*, the major challenges to pursuing any kind of livelihood are the lack of well-paved roads that connect to town markets, severe shortage of clean water supply both for animals and human beings, lack of fully functioning health services and absence of electric power.

4.3 Agriculture and food security in the district

Agriculture in rural Meskan is mixed and includes production of both crops and livestock. Farming is largely subsistence and farmers use oxen-driven ploughs. Maize and *enset* are the major staples in the district. However, *enset* (false banana), a perennial and drought-resistant crop, is virtually never the sole food source of households and is rarely the chief market item. Low productivity caused by bacterial wilt is a major constraint that discourages *enset* production. Crops such as *teff* (very fine-sized grain used as staple food grain in some parts of Ethiopia), wheat, barley, *zengada* (indigenous variety of sorghum), haricot bean, chickpeas are widely grown. In addition to these, red pepper, coffee, chat and fruits are grown as cash crops (SNNPR Livelihood Zone Report 2005; DPED 1998).

According to the Atlas of Ethiopian Livelihoods released in the year 2010, Meskan district comprises four livelihood zones. The study *kebeles* fall under the Gurage-Silte *enset* and *teff* livelihood zone which covers the central (*woina dega*) areas of Gurage and Siltie administrative zone (MoARD 2010). Figure 4.3 shows the agricultural calendar of this livelihood zone. As shown in the diagram, agriculture is rain dependent and is characterized by its bimodality: the *belg* rains last from March to May and the *kiremt*[1] rains from June to September. Most land preparation takes place before the start of the *belg* rains and continues until the beginning of the *kiremt* rains, since some *kiremt* rain dependent crops such as *teff* require the land to be repeatedly ploughed and thus are very labour demanding. Most crops are planted with the start of the rains. There are no *belg*-dependent crops, but maize is the most commonly grown crop during this rainy season. The *belg* rains are crucial to the availability of water for humans and livestock as well as for pasture. It is also important for the growth of perennial crops such as chat and coffee. Food purchases peak in the months running up to the start of the green maize harvest – the annual "hunger" season.

In most years the hunger season lasts for three months, from May, when the main season crops run out, until the end of July, when maize is mature enough for green consumption. Food deficits during this period are met through purchasing of food from the market. This is also a period when livestock sales are high, as households sell animals in order to obtain cash to purchase food. Livestock are also sold during religious holiday periods. The main dry harvest period is in November and December. *Enset*[2] can be harvested at any time, but usually in November and December. Malaria is worst during the rainy season, and particularly in May and October, affecting labour availability at household level during these important months in the agricultural calendar.

[1] They are also known as *meher* rains. The main cultivation season is dependent on the *meher (kiremt)* rains.

[2] The diagram shows the planting and harvesting periods of *enset*, however, *enset* takes 4-6 years to mature. Even though it is eaten before its full maturity, it is impossible to plant and harvest within the same year (MoARD 2010).

Figure 4.3 Agricultural calendar of the Gurage-Siltie enset and teff livelihood zone

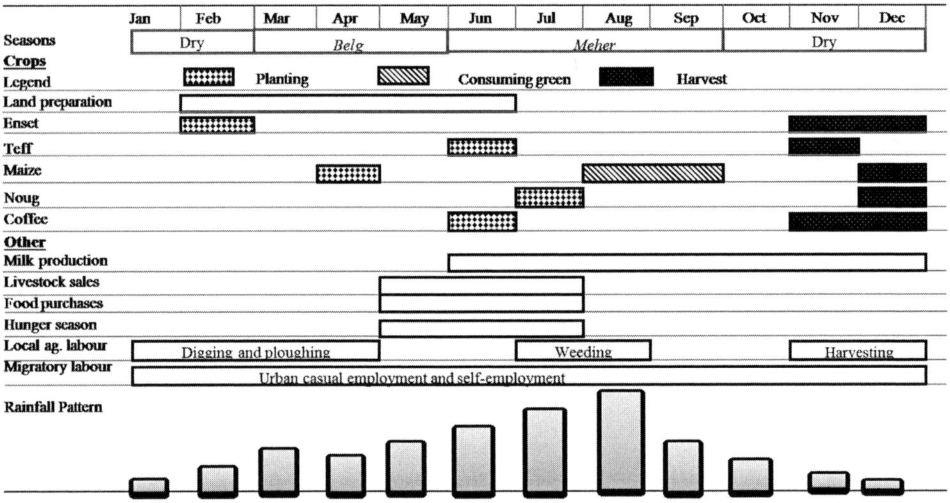

Source: Own modification after USAID 2005, 2

In Meskan district climatic hazards such as erratic rainfall in the form of delayed onset, early cessation or insufficient quantity or distribution of *belg* or *kiremit* rains are common. Occurrence of animal disease causes deaths and reduces milk production. These compound to reduce food production, making most inhabitants food insecure and malnourished (USAID 2005; CHF International 2005). According to an evaluation of crop production conducted by Meskan District Rural Development Office for the year 2008/2009, the planting of *belg* crops such as *zengada* and red pepper was constrained by the late onset and small amount of rain received. At district level, there has been a constant decrease of production of *belg* crops as lands covered by *belg* crops dwindle.[3] In the case of the *kiremt* rains, the report reveals high rainfall during times of crop growth and interruption of rain during time of seeding. High temperature for long periods caused a crop disease threatening the production of *teff* and wheat.

According to rainfall data gathered over nine years for Meskan district, the *belg* rain shows a slightly decreasing trend with higher variability over the years, whereas, the *kiremt* rains has shown an increasing trend yet, with less variability as compared to the *belg* rain (Figure 4.4).

[3] This is because farmers refrain from planting *belg* crops for fear of crop failure.

Figure 4.4 Seasonal rainfall distribution in Meskan district, 1994-2006

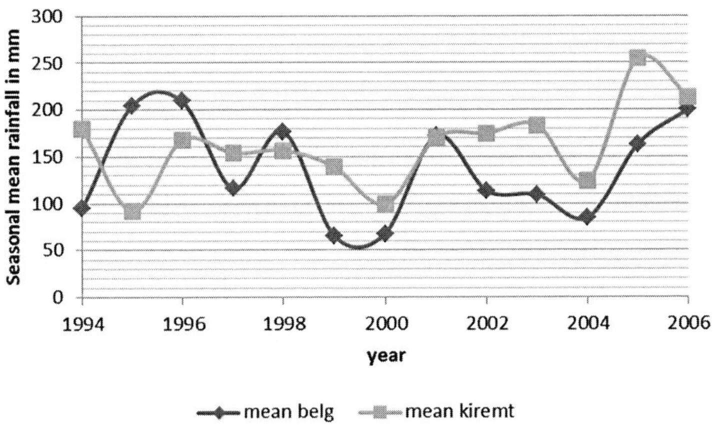

Source: Own design based on data obtained from Ethiopian Meteorological Agency, 2011

Table 4.1 shows the sequence of indicators of food security crises in the study area. It shows that failure and irregularity of seasonal rain is accompanied by various hazards that threaten the food security of households.

Table 4.1 Indicators of crises

Season	Month	Indicator
Dry	Jan.	High cereal prices during the harvest and immediate post-harvest period
Belg season	Feb.	
	March	Failure of *belg* rains
	April	Unusually severe outbreak of malaria
Dry	May	Unusually severe outbreak of malaria
Meher season	June	Late start of *kiremt* rains
	July	Uneven distribution and inadequate amount of rain-fall
	Aug	Uneven distribution and inadequate amount of rain-fall
	Sept	Delayed start of green maize harvest
	Oct	Unusually severe outbreak of malaria
Dry	Nov	
	Dec	High cereal prices during the harvest and immediate post-harvest period

Source: SNNPR Livelihood Zone Report 2005, 5

4.4 Food Security Programme in Meskan district

In Meskan district, various components of the food security programme have been implemented with the aim of assisting households to fill food gaps, prevent asset depletion and assist asset formation, at both household and community levels. Based on a report obtained from Meskan District Agriculture and Development Office (2011), the major types of interventions include the Productive Safety Net Programme (PSNP) consisting of direct support and community asset building "food for work" schemes as well as the Household Asset Building Programme (HARP).

As part of the government of Ethiopia's wider Food Security Programme, the Productive Safety Net Programme (PSNP) began in 15 highly food-insecure *kebeles* in Meskan district in the year 2005. By the year 2006, the programme served a total of 24 *kebeles*, covering 17 percent of households in the district who experience food shortages for periods from three to six months (February to August) each year (Meskan District Agriculture and Development Office 2011, 1).

These beneficiary households in the district are supported through direct food or cash transfer. The direct support programme targets households whose members are less able bodied due to age and those who do not have additional labour to participate in public works, households with inadequate family support, and those who are disabled. Pregnant and lactating mothers are also targeted by the direct support programme, starting from the fourth month of pregnancy till birth in the case of the former and for the first ten months after birth in the case of the latter.

The community asset building programme in Meskan district is another component part of the "food for work" scheme of the PSNP, involving nearly 83 percent of households in the district. Some of the major activities that are being undertaken include mobilizing able-bodied community members in activities such as soil and water conservation, water harvesting, reforestation and hillside closures, and rural infrastructure development such as road maintenance, building schools, health posts, farmers training centre, toilet construction, and so on. Between 2007 and 2010, a total of 316.7 km rural inner (connecting roads) were built, maintenance of 551 km rural inner connecting roads was carried out in the district. Moreover, six different level bridges were constructed. New schools were built, in addition to the renovation of 27 schools and 26 classrooms in various schools. A total of 6 farmers' training centres, 3 houses for development agents, 4 community health posts and 4 veterinary centres were built in the district.

The family package programme consists of asset building and/or income improving schemes. Around 3312 households were made beneficiaries of the family package programme in the district. It is meant to improve household income, fill food gaps, and encourage households to become self-sufficient on a long-term basis after graduating from the Productive Safety Net Programme (Meskan District Agriculture and Development Office 2011, 1).

4.5 The condition of women in the district

Women in Meskan district comprise slightly more than half of the total population. There are half as many female-headed households as male-headed households in the entire district. In this largely food insecure district, women's contribution as food providers and managers of households is tremendous. They play an active role in devising and carrying out survival strategies as they undertake their productive and reproductive roles both in the domestic and non-domestic spheres. Yet, poor women and their children face serious risks of various kinds emanating from malnutrition, poor sanitation, unhygienic housing conditions, etc. Diseases such as eye infection, goitre, elephantiasis and bacterial infections are rampant (Progynist 2002).

The deprivation of women in various aspects of their lives reflects their low social and economic position. For instance, with regard to school education, despite their high enrolment level at the primary and lower secondary levels, girls are largely absent at the higher secondary and tertiary levels (Meskan District Finance and Economic Development Office 2011). A large majority of women in the district are illiterate. The fertility rate among women of productive age is one of the highest in the country. Birth-related infant mortality is common, for women give birth to many children under poor maternal and child health services. Women are discouraged from participating in activities outside their traditional role in the domestic sphere, and thus are absent in decision-making processes in community and higher public spheres (Progynist 2002).

Rural women in the district are also victims of gender-biased local norms, societal views and perceptions, as well as customary practices and gender-based violence that are perpetuated on the pretext of preserving culture and tradition. Table 4.2 presents an assessment report for three districts of the SNNPR in the year 2004, of which Meskan was one among them. The study revealed around 17 top ranking harmful traditional practices (HTPs) and forms of violence which have a detrimental effect on women's well-being. It identified marital related rape, abduction and female genital mutilation as the most pervasive forms of violence against women, ranking from first to third respectively. Polygamy and early marriage are also relatively severe, taking the fourth and the fifth position in the district.[4]

HTPs and gender-based violence like polygamy, rape and circumcision and others have passed from one generation to the next. The small minority of the population that does not subscribe to these harmful cultural beliefs and practices are faced with a formidable challenge. Even though there have been significant improvements in the last two decades due to massive NGO intervention in some parts of the district, progress has been slowed in the last couple of years[5] and communities still practise them in various ways (Aster and Azeb 2004).

[4] HTPs such as dowry, polygamy, unequal property rights and gender biases in inheritance are also practiced in Amhara, Afar, Gambella and Oromiya regions of the country (WAO and WB 1998).

[5] This is partly caused by the absence of NGOs working on human right related issues due to the new NGO Law introduced in 2009 in the country.

Table 4.2 Frequency rank of violence types in Meskan, Mareko and Soddo Districts, SNNPR

No.	Type of violence	Rank
1	Rape	1
2	Abduction	2
3	Female genital mutilation (FGM)	3
4	Polygamy	4
5	Early marriage	5
6	Battering and psychological abuse	6
7	Unmatched/incompatible marriage	7
8	Wife inheritance	8
9	Unequal property rights	9
10	Women's inability to inherit land	9
11	Unequal educational opportunities for girls	11
12	Removal of nails (before marriage)	11
13	Purposeful transmittal of HIV/AIDS	11
14	Unwanted pregnancy	14
15	Forced single parenting	14
16	Gender based discrimination	14
17	To take away virginity	14

Source: Aster and Azeb 2004, 15

5 Research Methods

Following Hancock *et al.* (2007), a case study, as a research design, follows a particularistic and contextual research approach to provide an in-depth analysis of a single or small number of units. This case study was conducted in Meskan district with the aim of generating an area-specific knowledge through an in-depth understanding of the livelihoods of farm households. Meskan district is purposely selected considering factors such as the prevalence of food insecurity, the severity of gender related disadvantages and the author's prior field research experience in the district. Out of 40 administrative *kebeles* in the district, four sample *kebeles*, namely, Beresa, Dobo Tuto, Dirama and Semen Shershera were selected for the empirical investigation. Prior to the selection of *kebeles*, gender experts from a women empowerment non-governmental organization (NGO), which is active in Meskan district, were thoroughly consulted. Factors such as the severity of food insecurity, the prevalence of female headship, and the accessibility of transport to each of the study *kebeles* were considered in the selection process.[1]

The following six sub-sections shed light on the research methods employed for the study. The first section discusses the theoretical and methodological implications of the research approach used with regard to the problem under investigation. Then an attempt is made to give an overview of the important procedures followed during the period of the field work. The third and fourth sections deal with the method of data collection and the sampling design, with particular emphasis on household survey and qualitative inquiry techniques. The fifth section presents the methods of data analysis. Finally, the chapter winds up by presenting the main problems encountered during the field research.

5.1 Mixed research approach: integrating qualitative and quantitative research approaches

The qualitative research method is different from quantitative research in terms of its philosophical, epistemological and methodological perspectives. As summarized in Mack *et al.* (2005) quantitative and qualitative research approaches deviate from each other primarily in five major areas: general frame, analytical objectives, question format, data format, the forms of data they produce, and the degree of flexibility built into the study design.

While the debate continues regarding the pros and cons of qualitative and quantitative research methods, a mixed methods approach combining both qualitative and quantitative methods is enjoying increasing popularity in studies dealing with humans in general. As stated in Creswell (2003), this approach facilitates pragmatically based knowledge which is consequence-oriented, problem-centred and pluralistic by its nature. It allows more flexibility in that data can be collected in a simultaneous or sequential manner; the end result is qualitative data in the form of interviews or text data, as well as quantitative information. According to Greene *et al.* (1989, 259) the benefits of the mixed research approach include triangulation, which means *"convergence, corroboration, correspondence of results from the different methods"* and complementarity, which *"seeks elaboration, enhancement, illustration, clarification of the results from one method with the results from the other methods"*. Another benefit of the mixed methods approach is that it uses *"results from one method to help develop or inform the other method, where development is broadly construed to include sampling and implementation, as well as measurement decisions."* They also pointed out that mixed meth-

[1] The author has gathered adequate information on the overall situation of poverty in the district and the study *kebeles* through interviews held with gender experts working in Progynist, a secular, non-profit women empowerment NGO that has been active in the study district for more than 15 years. Moreover, various unpublished reports of the NGO were reviewed.

ods are helpful for initiation, which *"seeks the discovery of paradox and contradiction, new perspectives of frameworks, the recasting of questions or results from one method with questions or results from the other method"* and expansion, which *"seeks to extend the breadth and range of inquiry by using different methods for different inquiry components."*

The author believes that neither qualitative nor quantitative methods alone would be able to capture the complex and multi-dimensional nature of the livelihoods of the study population. Thus, a mixed methods approach was used to gather data on a wide range of topics. The information obtained through a mix of qualitative and quantitative data collections tools is used in a complementary manner, in that it helps to complement the strengths and compensate for the weaknesses of each method used. A mixed research approach was also a useful method in that it facilitated triangulation. For instance, in some cases focus group discussions (FGDs) proved damaging to the collaborative atmosphere among the discussants because of the confidential nature of the topics under discussion. Thus, these kinds of topics are addressed through interviews with knowledgeable key informants. In other cases, structured interviews through questionnaires proved less effective for collecting quantitative data on topics relating to matters such as use of agricultural inputs, yield, etc. To fill such gaps, additional data is obtained through key informants. Application of different data collection tools on the same issue enabled cross checking and thus was helpful in ensuring the quality and relevance of the data gathered. Triangulation was also made possible by asking focus group discussants the same questions that are included in the household survey questionnaires. Generally, a mixed methods approach helped to strengthen the validity of the findings, facilitated data gathering on a wide-range of topics, and thus provided an in-depth understanding of the subject under study.

5.2 Relevant procedures in the course of the field work

The last month of 2010 was devoted to preparation to travel to Ethiopia for field data collection. One of the preparations included the designing of checklists and guiding questions to be used for the various tools of qualitative inquiry. Thus, various secondary materials were reviewed as guiding sources for the preparation of the questions. Most notably, I benefited from reviewing the Meskan district livelihood profile (MoARD 2010) as well as unpublished documents gathered from a women empowerment NGO called Progynist that has long years of experience in the study area.

During the first month of my stay in Ethiopia, I established social and professional networks which made access to some of the support I needed easier. Contacts were made with the NGO[2] mentioned above, and with government offices at district as well as *kebele* level which are of relevance to my fieldwork. As a result, I was able to get official permission to carry out data collection in Meskan district. A field study was conducted in two different periods including the major field data collection in the year 2011 and a short field visit in the year 2012.

My acquaintance with the people working in the NGO made the process of permission easier for me, since the organization has long-established working relations with the district and *kebele* administration offices where permits are issued. The NGO office at Butajira town was an ideal spot for meetings with the enumerators and the research assistant throughout the field work. I was also offered a room to carry out group discussions, which created a very quiet and suitable environment to audio-tape the discussions. Furthermore, through the collaboration of the *kebele* officials, the staff members of the NGO participated in the process of recruiting enumerators, a research assistant and FGD participants. Eventually, I was able to get access to

[2] I established relations with some of the staff members of the NGO prior to the beginning of my PhD research.

enumerators who are health extension workers native to the study *kebeles*, with college degrees and experience of data collection. Due to their frequent contact with the study population, the enumerators faced no barriers in gaining admission to people's houses. This facilitated the accurate and timely completion of the survey. Figure 5.1 provides an overview of the major activities and procedures followed during the entire field work.

Figure 5.1 Summary of the field data collection procedures

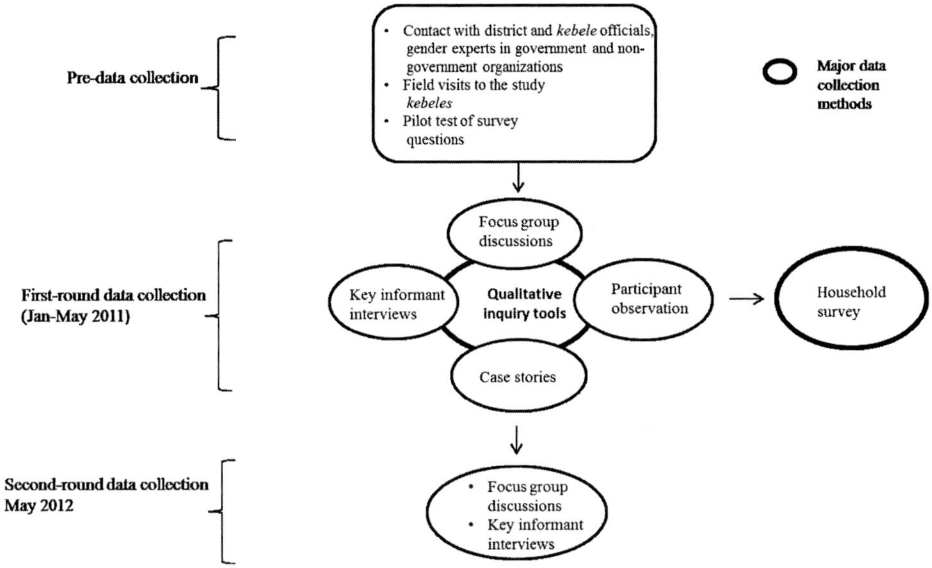

Source: Own design

5.3 Sampling design: Sampling procedure and sample size

For the household survey, both female-headed households (FHHs) and male headed households (MHHs) in Beresa, Dobo Tuto, Dirama, and Semen Shershera *kebele* constituted the sample frame. Out of the 2203 households in the four *kebeles*, a total of 372 households, or 17 percent of the total households, were taken as study samples.[3] A relatively large sample size was chosen for the sake of reducing the effect of unsuccessful questionnaires and ensuring the quality of the data. A stratified simple random sampling technique was applied to select representative samples from the two categories of MHHs and FHHs. In this case, FHHs are found to be smaller than MHHs. Since the aim of the study is to compare results between MHHs and FHHs, it was found necessary to over-sample FHHs in order to achieve statistical power and to demonstrate the validity of differences between the two groups. Therefore, disproportionate stratified random sampling was used to select representative samples from MHHs and FHHs. Considering the total number of female- and male-headed households, an equal number of sample households, i.e. 186 from each category, were taken as representative samples of the survey. Based on the total list of names of household heads in each *kebele,* a lottery technique was used to randomly select respondent households for the survey.

[3] The computer software Epi Info version 3.5.1 was employed in determining sample size.

Table 5.1 shows the total household size and the sample size by the gender of head of households in the four sample *kebeles*.

Table 5.1 The number of female- and male-headed households in the study *kebeles*

Kebeles	**FHHs**		**MHHs**		**Total**	
	Number of Households	Sample size	Number of Households	Sample size	Total number of FHHs & MHHs	Total sample size
Beresa	190	60	503	60	693	120
Dobo Tuto	76	40	346	40	422	80
Semen Shershera	157	43	371	43	528	86
Dirama	189	43	371	43	560	86
Total	612	186	1591	186	**2203**	**372**
Sample size (%)		30.4		11.7		**17**

Source: Administration offices of the respective *kebeles,* 2011

In the case of qualitative data, the purposive sampling method was employed to select respondents for focus group discussions and key informant interviews. Attributes such as age, sex, and social and economic status within the community were considered during the selection process. NGO workers and community members in positions of official authority were consulted in order to recruit potential participants in each *kebele.* People who were less accessible to the researcher but who could potentially provide input for the study were accessed through a snowball sampling method. This was made possible through the social networks of informants whom the researcher made prior contacts with.

5.4 Method of data collection

The study used both primary and secondary data sources. Primary data was collected using household survey and various qualitative inquiry techniques such as focus group discussions, key informant interviews and participatory observation. Secondary data were obtained by reviewing development programmes and policy documents, food security strategy papers, books, articles, proclamations as well as both published and unpublished reports.

5.4.1 Household survey

Both structured and semi-structured questionnaires were designed and used as data collection tools to generate quantified data (see Appendix I). Following the concept of Bernard (2011), semi-structured interviews were used to probe for more information based on the checklist of questions to guide the interviewee. Open-ended questions were included where the author had limited knowledge about the extent of the list of possible answers. This allowed the respondents to freely give detailed information. Structured questions were used for questions that are known to have limited responses. However, options such as "other (s), please specify" were included to give respondents an opportunity to provide more information.

Before the questionnaires were administered in the study area, the questions were translated into Amharic.[4] Then, the survey questions were pilot tested and necessary revisions were made to make each and every question applicable to the local environment. The pilot testing helped to re-define options for the structured interview. Following this, the survey was admin-

[4] Even though the study population is largely from the Gurage ethnic group and speaks the local Gurage language, Amharic is also widely spoken in all of the study *kebeles*.

istered for 30 consecutive days (1 March to 1 April 2011). The method of data collection applied for the household survey was a face-to-face interview. A total of 8 enumerators participated in the data collection, supervised by the author and one research assistant. After the household survey was completed, discussions were held with the enumerators on the challenges that were faced during the data collection and notes were taken.

5.4.2 Qualitative inquiry Tools

In order to generate qualitative data, several according techniques were applied, including focus group discussions, key informant interviews, participatory observation, and case stories. Focus group discussions were conducted in two different periods. The first one was from 8 to 16 February 2011. A total of two FGDs (each consisting of female heads and male heads separately) in each of the four *kebeles* were held making a total of 8 FGDs. In addition to this, one FGD were also held with mixed groups of female and male household heads in the year 2012. Generally, each focus group discussion involved eight participants, and lasted from two to two and half hours. To ensure group diversity, in the case of female heads, the participants included widowed, divorced and separated women at different age levels and from different locations within each *kebele*. The groups containing male heads were also organized to include men of different ages. Attempts were made to include individuals from different villages and with different roles in their respective communities. It was also important that one or two participants in each group were beneficiaries of the government food security programme (among both the women and the men). Figure 5.2 shows photos captured during FGDs with an exemplary men group (on the left hand side) and women group (in the right hand side) during the field work.

Figure 5.2 FGD sessions with female and male heads of households

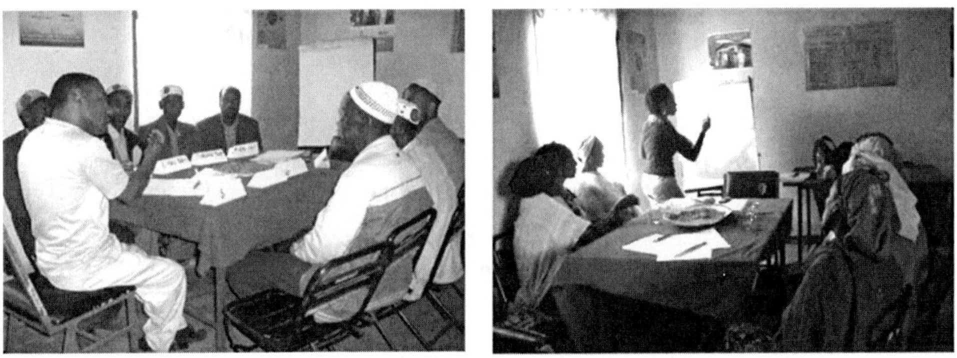

Source: Focus group discussions, February 2011

The group dynamics in each of the FGDs facilitated the collection of well-founded data on a wide range of topics. Substantial information was obtained about shared norms, and cultural and social practices in the study area. Group perceptions and opinions about the various norms, values and customary practices within the community contributed to a broad understanding of the most dominant local institutions and provided insights into the gender elements in their constitution. The variety of views within the groups also gave insight into the complexity of the mode of life of men and women in the study *kebeles*. Farmers' opinions and views on topical issues regarding commonly felt environmental, social and economic shocks and stresses on resources, and their coping strategies, helped to complement the quantitative data obtained through the household survey (see Appendix II).The opinions and views of beneficiaries regarding Safety Net and other food security interventions provided key information to assess the role of policy in filling food security gaps in female- and male-headed households. The FGD was also instrumental in capturing personal experiences and views of individuals, which facilitated the process of identifying potential key informants for further discussions.

In addition to FGDs, a total of 17 face-to-face key informant and expert interviews were conducted between January and April 2011. The interviewees included people with posts at *kebele* and district levels. Among these are Food Security experts, Development Agents (DAs), Gender experts, *Kebele* chairpersons and Community Leaders (see Appendix III). The interviews were essential tools for capturing the perspectives, feelings, opinions and experiences of individuals on topics that people might be reluctant to discuss in a group setting. They provided in-depth information by complementing the information obtained from FGDs. For instance, in-depth interviews with people assuming traditional positions within their community, including *idirs*[5] and local customary judiciary system helped to capture the historical dimensions of institutions and environmental and social and economic transformations in the study area (see Appendix IV). The data obtained from experts was used as input to analyse the existing policy gap through a comparative assessment of the perceptions and views of local actors and people who are implementing government policies.

Furthermore, in order to obtain an "insider" view of the physical, social, cultural and economic contexts in which the study participants live, participatory observation was conducted prior to the commencement of the actual data collection in the study *kebeles*. In addition, participatory observation was also conducted parallel to the other data collections. This helped to complement the information obtained from key informant interviews by reducing data inconsistency. Accordingly, informal talks were held with community members as they engaged in their daily routines in different places, for instance, around their homes, at water points, and in market areas. This helped to gather information about the nature of social interactions between men and women and the contexts within which such interactions occur, and to understand norms, events, people's behaviours and activities including what they do, how frequently and with whom. Meanwhile, field notes were taken carefully based on the observations made.

It is difficult to present a full and comprehensive picture of all the practical ways in which people live in their social environment by simply depending on the data collected through interviews. Yet, incorporating specific examples can offer a good chance to create a more general picture. In Table 5.2, an attempt is made to demonstrate the complex issues surrounding vulnerability, food security and coping strategies, using examples to address how and why questions, following the recommendation of Yin (2009, 8).

[5] Indigenous saving schemes to help families with money and funeral procedures

Table 5.2 Summary of research methods used

Objectives/Research questions	Data collection tool	Operationalization
What factors define the vulnerability contexts of FHHs and MHHs?	Secondary sources review, key informant interviews, expert interviews, case studies, FGDs	Livelihood profile of Meskan district, unpublished documents obtained from a women empowerment NGO that is active in the study *kebeles*, documents at district, review of a report on filed cases at district land administration office, interview with elderly religious leaders and *shimagiles*, gender experts at the district, case studies of women victims of polygamy and property claims by men, FGDs with female heads and male heads
Why are female-headed households more food insecure than male-headed households in the study area?	Participant observation FGDs, expert interviews and household survey	Walk through the study *kebeles*, observation of farm land, livestock possession, and utilization of farm inputs and public services such as water points, group discussion with FHHs, interview with development agents
What are the types of livelihood shocks experienced by households and how do female- and male-headed households in the study area cope?	Participant observation, FGDs, expert interviews, household survey	Short visits to markets on market days, visit to individual houses, and walk through the study *kebeles*, group discussion with FHHs and MHHs about common shocks and resource stresses and household and community level coping mechanisms
How are the food security programmes and gender policies addressing the gender gap in food security in the study area?	Policy document review, FGDs, key informant and expert interviews	National Development Programme papers, food security strategy papers, Ethiopian women policy document, the constitution, the family code and land proclamation , group discussion with male and female safety net and food security programme beneficiaries, expert interviews with district food security desk, development agents, gender office

Source: Own design

5.5 Method of data analysis

Qualitative data is mostly a collection of textual stories rather than numbers (Auerbach and Silverstein 2003, 24). The key messages from each of the sources are investigated from the texts generated during the qualitative data collection process and compared with others to identify differences and similarities and give meaning.

Accordingly, all audio-taped qualitative information gathered through FGDs, key informant interviews and participatory observation were transcribed and the texts used either as a sole form of information or as additional information to complement the quantitative information. Exemplary statements made by interviewees and FGD participants were quoted to support discussions. Simple description is used to discuss the findings of the qualitative data. Generally, since a mixed methods approach is used for this study, the findings are also presented in such a way that the qualitative data is used to enrich the findings of the quantitative analysis, as described in Corbetta (2003, 46). In the case of quantitative data analysis, the Statistical Package for Social Sciences (SPSS) software was used to process and analyse the quantitative data.

5.6 Problems encountered during the field research

My field experience was not free from problems. One of the challenges was related to the interference of *kebele* officials in the selection of FGD participants. Through my own informal discussions with the local people, I found out that this was due to two main reasons. The first one was their suspicion that the meeting would have political contents and their fear of the possible effects of political movement among farmers. The second reason was the misconception that the meetings would be associated with benefits such as per diem, as a result some *kebele* officials wanted to make sure that their relatives were included. Therefore, I had to explain the purpose of the group discussions and of the study in general so that the selection of participants could take place free of unfair and biased selection procedures by the officials. Regardless of these efforts, three of the participants in two *kebeles* during the FGDs were *kebele* officials.

The other problem during the field data collection was related to the availability of certain data. This problem was met at two levels. First, it was difficult to get gender-disaggregated data regarding the wealth profile of Safety Net beneficiaries in all of the study *kebeles*, so that it was not possible to assess the role of government interventions in closing the gender gap in the study *kebeles*. Second, the politicization of the food security issue in the country was a barrier to getting the right information and data on crop production, fertilizer use, etc., as people at the district level were reluctant to reveal information. The data collection took place at a time when serious meetings and trainings were being held between officials and between officials and farmers to implement government programmes in the study *kebeles*. This resulted in a number of failed meetings with officials at both district and *kebele* levels.

6 Demographic and socio-economic characteristics of the study population

The capabilities, assets (stores, resources, claims and access) and activities of households make up the livelihood concept. And the means of securing a living is made up of the interaction between the above component elements of livelihood (Chambers and Conway 1992). Factors such as education, age, availability of human labour and asset possessions are important indicator variables used to determine the kind of living poor people lead. Likewise, farm households' demographic and socio-economic characteristics are assumed to provide pertinent information in terms of understanding gender-based differences with regard to vulnerability, food security and coping strategies, which are the central focus of this thesis. Primarily depending on the data gathered through the household survey, interviews and focus group discussions (FGDs) in four of the study *kebeles*, namely Beresa, Dobo Tuto, Semen Shershera and Dirama, this chapter presents the findings of the study regarding some selected demographic and socio-economic indicator variables characterizing female- and male-headed households.

6.1 Female household headship

Like the other rural parts of Ethiopia, local norms in Meskan district assign due recognition to men as household heads. Men's status as household heads within the traditional family structure allows them to exercise major decisions around domestic and extra-domestic affairs. The following response from a widow to the question regarding her status in her household roughly demonstrates the existing biases regarding the public use of words like "male head" and "female head" within the study community. The respondent said: *"I am the 'Aba wera' (male head) for my household.*" The use of the word "male head" by this respondent can be interpreted in various ways. Firstly, the respondent wanted to emphasize that she manages her household in a similar manner to a male household head. Secondly, her choice of a comparative term to describe herself, "a male head", shows her preference for a normatively appropriate expression and widely used terminology which fits local cultural values. The term 'female head' is not understood in her community as a powerful position. All FHHs that are represented in the survey findings are, however, households which are independently managed by women. In other words, these women are exclusively responsible for all decisions regarding their household affairs. The nature of local norms and values and related topics are treated in detail in 7.2.

One of the most important social attributes of households in the study *kebeles* is captured by differentiating household heads on the basis of their marital status. The survey findings show that unlike male-headed households[1], female headship is quite diversified in the study area. These variations are captured through the household survey and presented in Figure 6.1.

[1] All of the survey participants in the men's group are married.

Figure 6.1 Percentage distribution of female-headed households by the cause of female headship

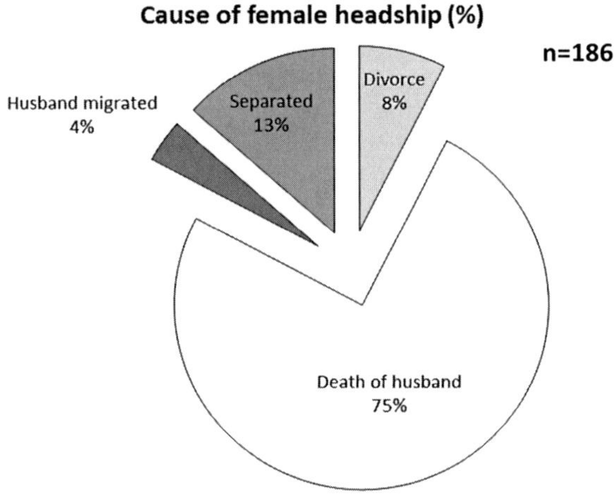

Source: Household survey, March to April 2011

The main reason for female headship in the study area is death of the husband, and widowhood represents around 75 percent. Even though the survey results do not identify the cause of death, informal discussions with elderly widowed women revealed that war-related causes are the primary factor causing early loss of male spouses. Separation makes up around 13 percent, in which all the respondents mentioned polygamous relationships[2] as the primary cause. Nearly 8 percent fall under the category of divorced female heads, out of which around 71 percent of the respondents relate the cause of their divorce to resource-related conflicts. Migration of a male spouse is the least mentioned reason for female headship, accounting for only about 4 percent. Thus, widowship, divorce and separation add together to make *de-jure* female headship the most dominant type of female headship among the study population.

6.2 Age of the household head

The survey findings revealed that a large segment of both male and female household heads fall within the age range of 20-60 years i.e. within the productive age category (Figure 6.2). Within this category, the figure for female heads is slightly lower (92 percent) than male heads (94 percent). This is in line with the aggregate national figure for the year 2012 in which slightly half of the total population is found to be within the productive age category (World Bank 2013). Among the study samples, a larger proportion of female heads are represented in the older age category of 41 to 50, whereas a larger proportion of male heads are within the age range of 31 to 40. Women are also disproportionately represented in the above-forty age category as compared to men. On average, female heads are older than male heads, the average age being 47 years and 43 years for female and male heads respectively. The World Bank statistical report on the average national life expectancy for the year 2011 shows 63 years for women and 61 years for men, so that women live longer than their counterparts.

[2] According to the 2011 Ethiopian demographic and health survey, the extent of polygamy has declined only slightly over the past six years, from 12 percent in 2005 to 11 percent in the year 2011. It also reports a higher prevalence among rural women (12 percent) than among urban women (5 percent). At the regional level, SNNPR and Benishangul-Gumuz together comprise 18 percent, next to Afar (22 percent) and Gambela (20 percent) (CSA and ICF International 2012, 60-61).

This strengthens our survey results which show a larger proportion of women in the older age category as compared to their male counterparts.

Figure 6.2 Percentage distribution of respondents' age by type of household

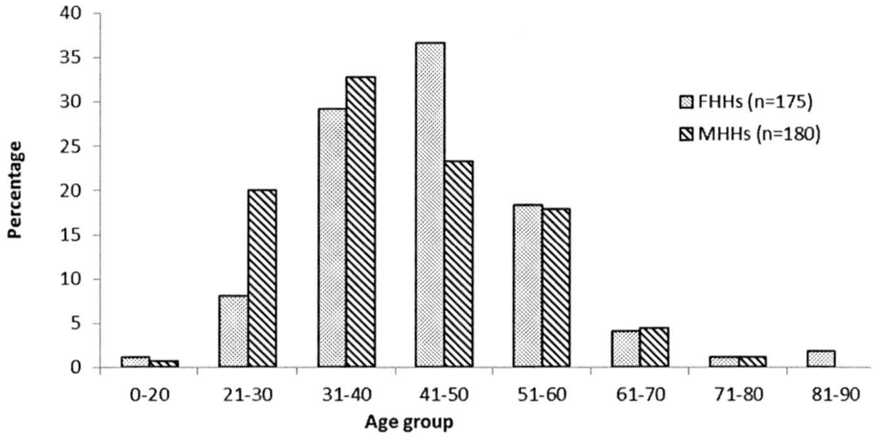

Source: Household survey, March to April 2011

In this thesis, the age of a household head is used as one of the most important explanatory demographic variables in describing the socio-economic characteristics of farm households. The age of a household head to a certain extent determines the availability of labour for various activities, both at household and community levels. For instance, the age of a woman can determine the number of children as well as the number of dependents within the household. A large proportion of female and male heads within the productive age category shows that the majority of household heads in the study area are able bodied, except in some extraordinary situations such as physical disability. Most importantly, in the context of this study, data on the age of household heads helps to trace the life path which farm households go through. In other words, it tells something about household heads' level of education, knowledge, perception and the nature of livelihood decisions that are important factors in the assessment of vulnerability, food security and coping strategies.

6.3 Level of education and religious background of the respondents

Education influences peoples' way of life. In the rural context, people with formal education are more open to employing new methods of production, or exploiting new livelihood options, or are able to engage independently in decision-making processes inside and outside their homes. Moreover, they are able to adopt various health practices to improve their family's health and well-being in general. Given these facts, it is not difficult to draw conclusions regarding the influence of the gender gap on the livelihood of female- and male-headed households. The empirical evidence in chapter eight helps to substantiate these conclusions.

As shown in Figure 6.3, there is marked gender gap in the education level of household heads. Illiteracy is higher among female heads of households, nearly 66 percent, which is more than double the percentage of illiterate men. Female heads with basic reading and writing skills comprise around 28 percent, whereas for male heads this figure is 19 percent. The adult literacy rate in Ethiopia is generally low, even though improvements have been made over the last decade. For instance, for the year 2011, adult literacy at the national level grew to 61 percent from 30 percent in the year 2005 (World Bank 2013). Moreover, educational attainment is much higher among the urban population than among the rural population. For example, in the year 2011, 28 percent of females and 15 percent of males in urban areas had no education, compared with 58 percent of females and 44 percent of males in rural areas (CSA and ICF International 2012, 27). Therefore, the empirical findings on literacy in the study area do not deviate from the statistical figures at the national and regional levels. In addition to this, the survey shows a marked gender gap in terms of primary level education with only 6 percent of women as compared to 44 percent of men. In other words, the proportion of male household heads with primary level education is eight times greater than that of female heads. Generally, the findings show a visible gender gap in terms of level of education for farm households in the study area. This is also represented in a high chi-squared figure (x^2 =89.2), significant at the 1 percent level.

Figure 6.3 Percentage distribution of respondents' education level by type of household

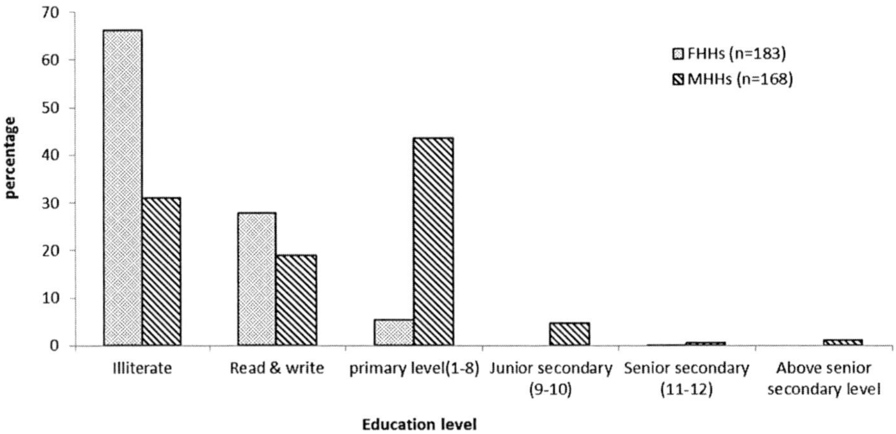

Source: Household survey, March to April 2011

At aggregate level, Islam dominates with nearly 67 percent of households, followed by Orthodox Christianity (26 percent) and Protestantism which represents only 7 percent of the total respondents. Islam is the dominant form of religion in both female- and male-headed households, followed by Orthodox and Protestant Christianity. Disaggregated along gender line, Muslim female heads represent a greater proportion than their counterparts (Figure 6.4).

Figure 6.4 Percentage distribution of respondents by religion and by type of households

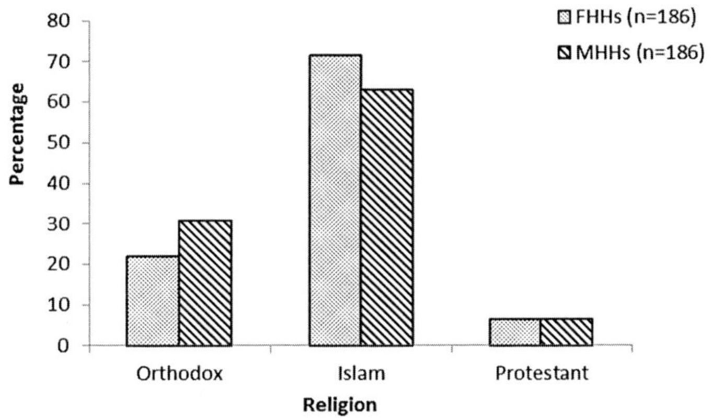

Source: Household survey, March to April 2011

Moreover, the survey results on the religious affiliation of respondents vary across *kebeles*. Semen Shershera is heavily dominated by Muslims, who account for 93 percent of the respondents. Relatively large numbers of Protestants are reported in Dirama *kebele*, which is probably due to the close contact of its residents with Protestant-affiliated NGOs that were active in the area (Figure 6.5).

Figure 6.5 Percentage distribution of respondents' religion by *kebele*

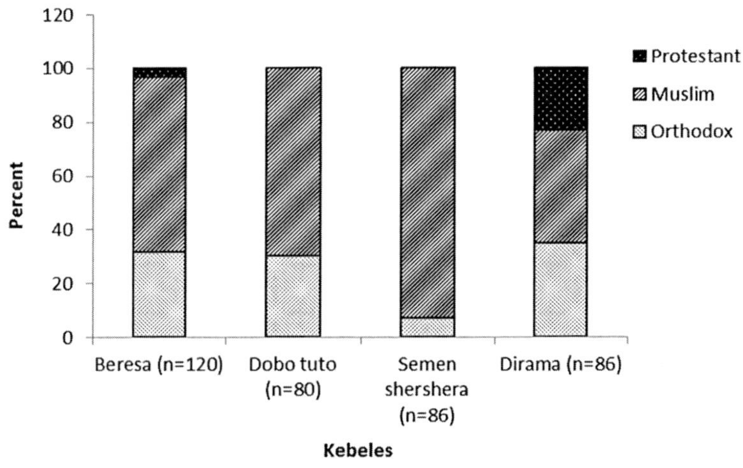

Source: Household survey, March to April 2011

In some instances, religion in the study area is intertwined with local norms, cultural and customary practices shared by people with different religious backgrounds. For instance, a customary practice such as polygamy is common among the Muslims and Orthodox Christians under the umbrella of religion and culture.

Religious and cultural differences, however, have little effect on the harmony among the people. Long-standing labour-sharing arrangements[3], and burial associations such as *Idir* are common forums for community members, regardless of the religious background of individuals. A number of other customary practices and shared societal norms have also contributed to peaceful relations among the people, regardless of religious differences.

6.4 Household size

The survey result in Figure 6.6 shows that household size in the study area covers a wide range, from one to fifteen persons per household. A large proportion of female- and male-headed households fall under the category of households with four to six persons. This category accounts for 59 percent of FHHs and 44 percent of MHHs. However, female-headed households are disproportionately represented in the lowest category of 1-3 household members. The figure in this case is more than double as compared to male-headed households. The average household size in FHHs, 4.7 persons against 6.6 in MHHs, shows that FHHs have fewer household members than their counterparts. Compared to the average rural household size in Ethiopia, which stood at 4.9 in the year 2011 (CSA and ICF International 2012, 21)[4], it can be asserted that the study area falls under those rural parts of the country with a high fertility rate. Moreover, the lower average family size recorded for female in comparison with male-headed households shows a similarity with previous studies in Amhara and Tigray regions (see Dejene 1994; Yared 1995; Mirutse *et al.* 2006).

Figure 6.6 Percentage distribution of respondents by household size and type of household

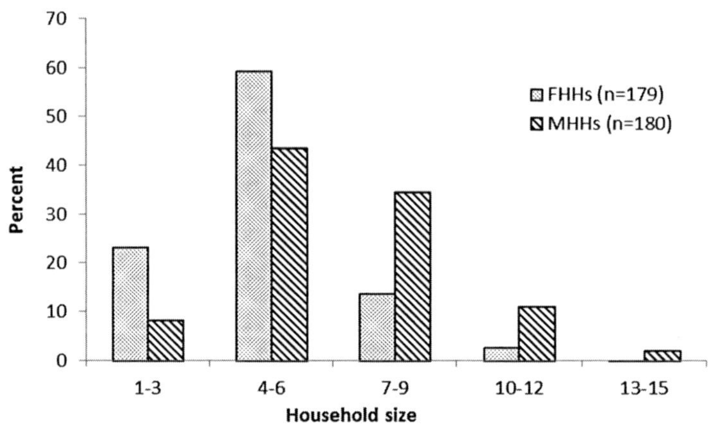

Source: Household survey, March to April 2011

The survey showed that a large number of households in both cases has one, or more than one, of their children living outside the home. In this case, 67 percent of female-headed households reported absentee children, against 47 percent of MHHs. Children in both male- and female-headed households leave home for two major reasons. Among those households where absence of children is reported, 53 percent of FHHs mentioned economic independence

[3] Some of these labour-sharing arrangements are explained in chapters eight and nine

[4] The same source also revealed that national average household size in 2011 was 4.6 persons, which is slightly lower than the average of 5 persons per household reported in 2005. Compared with rural areas, urban areas have a lower average household size, which the data for the same year shows (3.7 persons per household).

as a reason, whereas only 39 percent in MHHs attributed it to economic independence.[5] The second most reported reason for the absence of children is education, which was given as a reason by nearly 26 and 28 percent of female- and male-headed households respectively (Figure 6.7). There are two ways in which children may leave their parents' house because of education. Firstly, it is common practice for a group of students attending secondary school in Butajira town to rent a room and live temporarily together. Secondly, children may leave their parents' house to attending university in a distant location.

Figure 6.7 Percentage distribution of respondents' reasons for absence of children by type of household

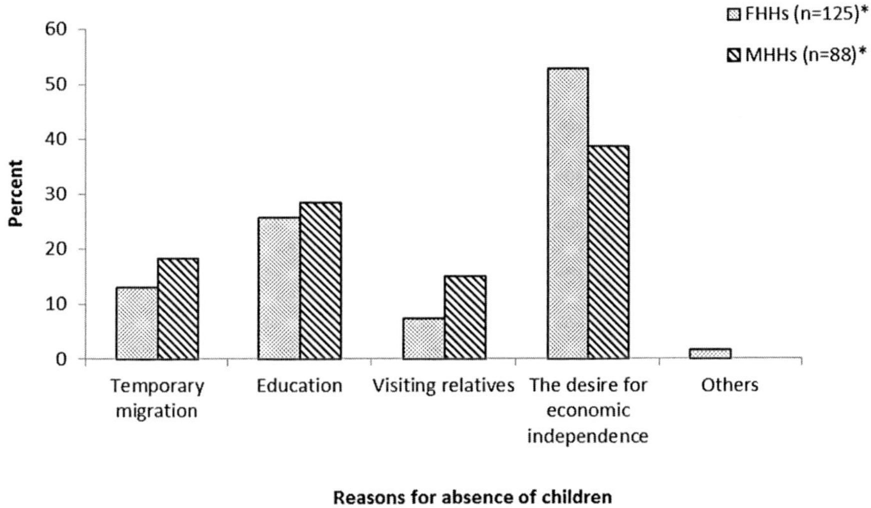

* Only those households who reported absentee children are included

Source: Household survey, March to April 2011

The availability of human labour at household level is an important factor that affects various on-farm and off-farm activities. For instance, female heads of households who have grown-up children are more likely to benefit from their children's labour (except in situations where the children are absent or do not make regular visits). Even though it is difficult to determine the number of dependent and independent members within FHHs from the data presented above, the age range within which most of them are found, the smaller average family size, and the greater number of absent children in female-headed households indicate that they are relatively more labour-constrained than MHHs.

6.5 Sources of livelihood

Most of the survey respondents identified farming as their major occupation. In addition to agriculture, households also engage in various non-agricultural activities. Generally, female- and male-headed households employ both on-farm and off-farm livelihood strategies to achieve food security and reduce vulnerability to livelihood risks.

[5] These are mostly households where children travel to towns and bigger cities as well as Middle East countries (a growing phenomenon at the time of data collection) in search for jobs.

Agricultural activities include the production of crops, vegetable and fruits and the growing of trees, as well as the rearing of livestock. The latter include the rearing of cattle, goats, sheep and poultry. Table 6.1 presents the various on-farm activities carried out by female and male headed households in the study area.

Table 6.1 Types of on-farm activities by type of household

Type of on-farm activities	**Type of Household**				**Total**	
	FHHs		**MHHs**			
	№	%	№	%	№	%
Crop production	127	71.3	103	56.3	230	63.7
Gardening	14	7.9	-	-	14	3.9
Crop production + gardening	8	4.5	20	10.9	28	7.8
Crop production + gardening + poultry	1	0.6	2	1.1	3	0.8
Crop production + gardening + planting trees for sale + poultry	-	-	5	2.7	5	1.4
Crop production + livestock rearing excluding poultry	28	15.7	52	28.4	80	22.2
Crop production + gardening+ planting trees for sale + livestock rearing	-	-	1	0.5	1	0.3
Total*	178	100	183	100	361	100

* The computed numbers exclude those households who do not take part in agricultural activities.
∑ Crop production FHHs= 164 (92.1 percent), MHHs= 183 (100 percent)
∑ gardening FHHs= 23 (12.9 percent), MHHs= 28 (15.3)

Source: Household survey, March to April 2011

The gender-disaggregated figures show that the majority of both male- and female-headed households are involved only in one type of agricultural activity. The available data shows that crops are cultivated by 92 percent of female-headed households and all of male-headed households. Next to crop production, the second largest category in both female- and male-headed households is gardening. This practice is pursued by a relatively larger proportion of MHHs (15 percent) as compared to FHHs (13 percent). Gardening is used as a sole option for agricultural production by only about 8 percent of FHHs, whereas in MHHs gardening is practised only in combination with crop production. Comparing the two groups, it stands out that dependence on one type of crop is more a livelihood characteristic of FHHs than MHHs.

In terms of the type of crops produced, maize is a staple grain in the study area. Moreover, grains such as *zengada*, *teff* and wheat are also grown. In addition to these, vegetables such as coffee, *chat* (a shrub whose leaves, when chewed, provide a stimulant drug), pepper, haricot beans, beans as well as fruits such as avocado, papaya, *kazmir* and mango are produced to some degree in the study *kebeles*. Among these, chat and coffee are preferred for their relatively good economic returns. Households grow trees both for home consumption, as sources of fuel wood and for market purposes. Eucalyptus is the most common tree variety in all of the study *kebeles*.[6]

[6] Growing trees and the production of young wood for sale is one means by which households cope with food shortage seasons in the study *kebeles*. Tree planting can take place on individual or communal lands. For in-

In the case of domestic animal production, the possession of domestic animals such as oxen and cows is very limited in all of the study *kebeles*. Households in general rarely own above a pair of oxen, which are primarily used for ploughing purposes. Compared to MHHs, oxen possession is extremely low among FHHs (Quantitative data on oxen possession are presented in 8.2.2). Domestic animal fattening is usually limited to farm households that are assisted by external government interventions and a few well-off farmers in the study *kebeles*. FHHs are more active in poultry production than their male counterparts. Both traditional and modern poultry production is practised by farm households in the study *kebeles*.

Off- farm activities

The mobility and entrepreneurial culture of the Gurage is well documented in the ethnographic studies of Worku (1996), in which the population is depicted as one of the most mobile ethnic group in terms of long-distance trade in Ethiopia. The study population is no exception. The qualitative information gathered from in-depth interviews with local elders reveals that households in the study *kebeles* mostly consist of sedentary farmers who are partly dependent on off-farm options for their livelihoods. FGDs also revealed that short-distance mobility for jobs outside agriculture is a common form of livelihood activity. Young and adult men are involved in wage labour in towns like Butajira, while women are largely petty traders. Women mainly engage in short-distance petty trading, which involves exchange of grains and other consumable commodities, usually in the neighbouring *kebeles*. Most of the off-farm activities also have a seasonal character. Their intensity becomes higher during the dry season or slack season than in the rainy season when agricultural activities are important and mobility is hampered due to poor road infrastructure.

A highly diversified pattern of non-agricultural activities is found among the study population. Households either take up one type of activity or combine two or more types of activities as a source of income generation. Table 6.2 presents a gender disaggregated summary of the various non-agricultural activities.

---

stance, in Semen Shershera, much of the tree planting activity is carried out on communal lands and the wood is used for building schools and farmers training centres, or other public services.

Table 6.2 Type of off-farm activities by type of household

Type of off-farm income-generating activities	Type of Household				Total	
	FHHs		MHHs			
	№	%	№	%	№	%
Selling of grain (petty trading)	25	16.1	51	29	76	22.9
Safety-net food-for-work programme	42	27.1	21	11.9	63	19.3
Petty trading of fuel wood	9	5.8	6	3.4	15	4.5
Wage labour in town	2	1.3	3	1.7	5	1.5
Working on individuals' farm land	1	0.6	3	1.7	4	1.2
SN + wage labour in town + on individuals' farmland	7	4.5	13	7.4	20	6
Wage labour in town + on individuals' farmland	-	-	10	5.7	10	3
SN* + selling grains (petty trading)	36	23.2	18	10.2	54	16.3
SN + petty trading of grain + wage labour	3	1.9	13	7.4	16	4.8
SN + petty trading of fuel wood	19	12.3	13	7 .4	32	9.7
SN + petty trading of fuel wood + grain selling	2	1.4	3	1.7	5	1.5
Selling of traditional drink + petty trading of grain	-	-	3	1.7	3	0.9
Not involved in off-farm activities	9	5.8	19	10.8	28	8.5
Total**	155	100	176	100	331	100

* SN= safety-net

** The computed numbers exclude missed values

∑ SN in FHHs=109 (70.3 percent), MHHs= 81(46 percent)

∑petty trading in FHHs= 94 (60.6 percent), MHHs=107 (60.8 percent)

∑wage labour in FHHs=13 (8.4 percent), MHHs=42 (23.9 percent)

Source: Household survey, March to April 2011

The largest proportion of FHHs (about 70 percent) identified the food-for-work scheme of the safety-net program as one form of off-farm activity. In MHHs, on the other hand, the safety-net program is identified as their main off-farm activity by only 46 percent. Petty trading is the second largest undertaking in FHHs accounting to 61 percent. Generally, excluding the safety-net programme and considering those who are able to engage in two or more types of activity, the findings show that FHHs represent a narrower spectrum of diversification than MHHs.

6.6 Local perception of wealth characteristics of households

A participatory wealth ranking exercise was conducted in the course of data collection, in order to characterize the respondent households based on their wealth status. FGDs held with men and women are able to capture farmers' perceptions of wealth measurement criteria that are widely shared by the study population. Variables such as number of oxen possessed, food self-sufficiency, size of farmland, the nature of involvement in the market, type of housing and degree of social reputation are mentioned as indicators of the wealth and well-being of households. Combining farmer's perceptions with the responses obtained from *kebele* officers and agricultural experts, possession of oxen, size of landholding and food self-sufficiency are found to be the most important indicators of a household's wealth status. Based on these criteria, the households in the study area are categorized as well off, poor, the poorest of the poor (destitute). The responses obtained from interviews and discussions are summarized in Table 6.3.

Table 6.3 Wealth indicators based on respondents' perceptions

Wealth rank	Wealth Indicators		
	Oxen possession	**Land size**	**Food sufficiency**
Well-off	≥ 2	≥ 1 hectare	Self-sufficient throughout the year
Poor	1 oxen	0.5-1 hectare	Up to 6 months food gap
The poorest of the poor	No oxen	0-0.5 hectare	> 6 months food gap

Source: Focus group discussions, February 2011

Well-off: Both MHHs and FHHs described well-off farmers as those who possess more than one pair of oxen[7]. They own larger plots of land (more than one hectare). They are also self-sufficient in terms of providing food for their household members. They are even able to produce surplus for the market. In this regard, they are involved in capital-intensive activities like livestock and grain trade. Perhaps a more refined view was given by the women's group, in which the discussants unanimously agreed on a statement by one female participant:

"A well-off farmer is not only able to feed his family three times a day throughout the year, he is also able to lend to others in bad times." [8]

Well-off families can usually afford to live in a house with a corrugated-iron roof, and are able to separate their shelter from that of their domestic animals. Such households generally have a good social reputation in their community, due to their decision-making and leadership skills in their homes, and their active participation in the social and political life of the community. For instance, they care about the health and education of their children. They play an important political and developmental role in their community by connecting the local people to various development programmes. In other words, they understand and facilitate implementation of various development projects. They are concerned about community resources and work hard for the peace and stability of their neighbours and community members.

Poor: Both female and male respondents shared more or less similar opinions regarding the characteristics of poor households. These households face food gaps that last up to 6 months and are hardly able to produce surplus for the market. They own less than a pair of oxen and usually enter into labour-sharing arrangements such as share-cropping, since they lack the necessary resources to produce independently on their farmlands. They live in traditional *tukuls* (thatched houses) and may or may not separate their shelter from that of their cattle. Most of these households depend on government support.

---

[7] Similar criteria are also used by agricultural experts to distinguish 'model farmers' in the study *kebeles*. However, the local people perceive this as a rough generalization which can be quite misleading. For instance, according to their perception, the title 'model farmer' is sometimes wrongly given to those whose livelihood primarily depends on external assistance.

[8] The fact that the women described a well-off household in terms of the ability to lend to others is a reflection of their experience of lending and borrowing with well-off relatives and neighbours in their community.

***The poorest of the poor*:** According to the information obtained from both men and women groups, the poorest of the poor in the third wealth category are chronically food-insecure groups who are unable to cover their food demand for most of the year. These households predominantly depend on off-farm activities, such as wage labour in towns, as a major source of income. They may own marginal land or may not own any piece of farmland at all. They often engage in share-cropping arrangements. They live in ramshackle huts that are vulnerable to floods and other environmental catastrophes.

The gender disaggregated survey results show that FHHs are disproportionately represented in the poorest of the poor and well-off categories. Figure 6.8 shows nearly 52 percent of FHHs in the poorest of the poor category and 14 percent in the well-off category, as compared to around 30 percent and 32 percent of MHHs respectively.

Figure 6.8 Wealth rank of respondents by type of household

Source: Household survey, March to April 2011

The distribution by sample *kebele*s shows that the largest proportion of respondents (nearly 50 percent) in Semen Shershera *kebele* are in the poorest of the poor category, followed by Beresa (43 percent). Dobo Tuto *kebele* shows a more balanced distribution in that it has a large number of respondent households in the well-off category (nearly 28 percent), with nearly 44 percent in the poor category and only 29 percent in the poorest of the poor category.

7 Vulnerability context and trends

There is ample literature that shows the close link between food insecurity and undesirable/vulnerable livelihoods (Davies 1996; Carney 1998; Scoones 1998; Ellis 2000). Thus, it is relevant to examine the nature of the vulnerability that reinforces negative livelihood outcomes such as food insecurity in the study area. This chapter starts off with the presumption that in the face of institutional pluralism, informal institutions largely shape the vulnerability context of female-headed households. The chapter is divided into two major sub-topics. The first section presents the findings of the study regarding local views or actors interpretation of vulnerable livelihoods. In doing so, the study opts to shed light on some important social and economic transformations against which the vulnerability of various groups of rural households can be assessed. Emphasis is given to the effects of these socio-economic shifts on the vulnerability of female- and male-headed households. This section also raises an important question: how do gender power relations relate to the different levels of vulnerability among the study population? The second section attempts to answer this question through an in-depth investigation of informal institutions in the study area.

7.1 Households' own perceptions of vulnerability

"Only people in local contexts know what is really happening on the ground! They are the ones who experience the hailstorms which destroy the buds of their beans and thus their produce...they are the ones who struggle to find their strategies of how to adapt to the recent stress."

Tröger *et al.* 2011, 9

For social and livelihood-related studies such as this thesis, this statement sends a very clear message about the importance of putting people, who are the study subjects, at the centre of the process of knowledge gathering. This is also one of the central themes of both the SLF and the 'theory of structuration' by which this study is framed.

Yet, although local people do not use the concept of "vulnerability" to describe their worsening situation, they feel the stress, face difficulties, talk about "risks", and make risk-taking and risk-avoiding decisions (Heijmans 2001, 6). Accordingly, farmers in the study area describe vulnerability in terms of the severity and frequency of environmental, social, economic and cultural shifts that they have witnessed in their village over the years. Key areas where change has been significantly felt are identified from the responses gathered through FGDs and key informant interviews with the local farmers in the study *kebeles*. These include increasing uncertainty of rainfall, livelihood practices, land use, resource possession, increasing stress on available labour and time, as well as changing trends in communal or social life.

Against the above backdrop, it is crucial to note that the primary aim of this section is limited to painting a picture of the existing environmental and social transformations in the study area, using farmers' own interpretation of their local context. The author's emphasis is given in certain areas where field observations are found relevant to strengthen the response of the respondents. Thus, it is by no means the objective of this section to present detailed statistical findings on each subject presented below. However, supporting empirical findings on each point are included in the subsequent chapters in areas where it is needed.

Irregularity and variability of Belg rains

All of the study *kebeles* are not new to drought events. Yet, participants in the FGDs expressed their worry that the frequency of drought years has increased lately. People have experienced the late appearance, small amount and early interruption of the *belg* rains, which are crucial for the production of short-term crops. Rainfall records over the last ten years in the study district indicate that there has been a drastic change in the amount of rain received in the district. A considerable variability in the *belg* rains can also be observed (see Figure 4.4). Likewise, a study on climate-change impact patterns at two exemplary sites in the Rift Valley region, where the study site is located, has identified shortage, delay or irregularity of the *belg* rains and increasing temperature as typical characteristic features of the region (Tröger *et al.* 2011).

Disruption of the local agricultural calendar

Farming in the study area is entirely dependent on rainfall. All of the four study *kebeles* have a similar pattern of rain throughout the year. As already described in section 4.3, there are generally two major growing seasons for crops. These are *belg* (the small rainy season, critical for fast-growing crops such as maize) and *kiremt* (the main rainy season). Based on these rainy seasons, households follow their own agricultural calendar, which is adapted to the rainfall pattern. The beginning and end of the rains influences households' decisions regarding the kind of agricultural activities to be carried out, the type of crops to grow and the kind of labour arrangements required for certain agricultural tasks. As the rainy season approaches, before its onset, households start to prepare their land for planting, so that they may start planting as soon as the rain starts. However, given the unpredictable nature of rain in the last few years, respondents often said that they have been forced to make some time shifts in carrying out their usual activities, such as preparation of the land, ploughing, and other related tasks. The local people perceive these shifts in the agricultural calendar as threats to their livelihoods. Participants of an FGD in Beresa *kebele* unanimously agreed with a statement made by one female head of household: *"This time last year we had already ploughed our land, but this year we have not yet started, because we are waiting for the rain to come."*[1]

Growing dependence on markets

The changes in weather patterns and the growing risks associated with depending on agriculture as the sole source of livelihood led to more dependence on food purchases and exchange on the market. As an elderly informant explained:

"The land these days has become old. We no longer harvest what we saw. Crop yield used to not only feed our household, but also used to be sold on the market for additional income. Today, our harvest is not even enough to feed the whole family. As a result, farmers have become purchasers of grain from the market."

An elderly farmer in Dirama *kebele*

By describing farm lands as "old", the respondent wanted to stress that the same land has been used for generations. Through in-depth discussion with the respondent, two important structural transformations became evident which are indirectly implied in the statement above. The first one is the fading away of age-old land management practices, which has led to poor productivity of farm lands in the study area. For instance, local conservation methods such as fallowing and crop rotation have been gradually abandoned.[2] The second form of transfor-

[1] The data was collected during the month of April which is the *belg* growing season in the study area. Farmers in all of the study *kebeles* had already noticed the delay of the rain and were unable to start ploughing their land.

[2] Factors such as high population pressure and increasing demand for agricultural land are common factors that have worsened the situation. In addition, information obtained from female FGD participants in Beresa *kebele* shows that the loss of soil fertility has been worsened due to the unregulated application of fertilizer. This is the

mation is the growing dependence of farm households on food purchases from the market to meet the household food demand (on the role of markets see 9.2.5).

Stress in labour and time allocation at the household level

Both female and male household-head respondents reported growing pressure in terms of the allocation of labour and time for tasks within and outside the household. The stress is primarily related to the decreasing role of child labour[3] both in MHHs in FHHs. Given the relatively laborious and time-consuming nature of agricultural tasks, the situation is worse in FHHs since they are doubly constrained due to the absence of male spouses. In such cases, women shoulder much of the responsibility for carrying out household chores and agricultural routines, which leads to their heavy dependence on male relatives and neighbours for agricultural activities such as ploughing (for detailed discussion see 8.2.3 and 9.2.2)

Disintegration of mutual and reciprocal social support systems

Social ties based on mutual or reciprocal social relationships play an important role in the daily routine and livelihood practices of men and women in the study area. This kind of social capital, that is built on local norms of sharing and reciprocal support, is particularly critical in times of need. The support includes lending money, giving moral support and advice when a family member is ill, or giving food or other items as a way of helping a household. Yet, the survival and proper functioning of some of these local institutional practices as long established coping mechanims have come under threat along with the externally changing economic forces. For instance, female respondents have noticed cuts in the frequency of coffee ceremony gatherings[4] (from three to four times a day to once or twice a week) due to the increasing price of coffee on the market. Some respondents revealed that they have ceased attending coffee ceremonies in their neighbourhood as they are unable to invite the others back. Similarly, the local people have gradually become unwilling to join labour-sharing arrangements such as *debo* due to the escalating value of labour[5].

The instability of market prices for non-food commodities has also contributed to the gradual erosion of community-level cooperation. The following story told by a male farmer in Beresa *kebele* can serve as a good example to show the impact of market price instability on community-level cooperation in the study area:

> *"Only last year, I got a title of recognition from my kebele administration office as a 'model farmer' as the best performing farmer in my kebele. I even used to lend my kebele up to 15000 Ethiopian birr to win against other kebeles so that my kebele gets priority during fertilizer distribution. But this year, not only was I unable to do that, I failed to pay for myself."*

The economic downturn experienced at the individual level, as illustrated by the above story, and the subsequent negative impact on the community level, has fractured social and economic structures. This might also be translated as a factor that creates a discouraging environment

---

case on lands owned by poor female-headed households since they are the most financially constrained groups without access to adequate inputs.

[3] Long distance out-migration of children for higher education and job opportunities has exacerbated the situation.

[4] Coffee ceremonies are important informal social forums for rural people in Ethiopia. These are an opportunity for people to share information and exchange ideas about private and community affairs.

[5] People no longer consider it worth working on their neighbours' farm land only for food, because of the growing importance of money in their livelihoods.

for social and economic cooperation between the well-off and the poor, with a long-term impact on community solidarity in the study area.

7.2 Vulnerability context: a closer look at informal institutions through a gender lens

Wisner *et al.* (1994) asserts that unequal power relations create vulnerability and uncertainties to local people and government policies and programs play an important role in the process. In section 3.2.1 of chapter three attempts are made to shade light on the role of the state and gender equality in Ethiopia. It is also indicated that institutional plurality is the prevailing institutional set-up in the country[6]. This section presents the gendered aspect of livelihood vulnerability through an in-depth investigation of informal institutions in the study area.

In the study area, local institutions such as marriage and family, gender roles and customary judiciary systems, and a number of other customary practices are some of the most important institutional manifestations that are predominantly built on patriarchal norms and values and unequal gender power relations. In the following parts of this section, each of them are explored in terms of their constituent elements, such as deeply rooted local norms, perceptions, beliefs and customary practices. I will attempt to shed light on some of the structural transformations that these institutions have gone through in time, stressing the dynamic nature of gender power relations. How the unequal gender power relations within these institutions relate to the unique vulnerability of women is also explored through a gender and livelihood lens. The overall aim is to show how the structuration process plays out in the actual field through mutual interaction between actors' agency and the existing institutional structures and processes within the study context.

7.2.1 Traditional marriage and family

Marriage is a key social institution, the foundation of family and society. Family is a central nucleus and the first organized institution where change can be initiated, nurtured and reproduced. In the context of rural Ethiopia, the family is an important institution in which gender roles and the behaviour of individual male and female household members are shaped.[7] Thus the role of marriage and family institutions is central for the maintenance of societal life.

The Ethiopian constitution recognizes marriage either through state or local traditional channels. The latter refers to either religious or custom-based marriage. Even though marriage through the state channel has become more accessible lately, traditional marriage is still a common practice among the study population. Using empirical findings gathered through FGDs and key informant interviews, the discussion below brings to light four major areas within the traditional marriage and family institution that are relevant to an understanding of livelihood vulnerability in relation to gender. Particular emphasis is given to institutional features such as patrilocality, values attached to the institution of marriage and family, perception of the "desirable traits of a married woman", and harmful traditional practices (HTPs).

Patrilocality

Patrilocality, an arrangement which forces women to move to their husbands' and his family's village upon marriage, is one of the local customary practices in the study *kebeles*. In the context of the investigation in this thesis, an in-depth assessment of the nature of patrilocal mar-

---

[6] Refer to section 3.2 for reading on formal institutions to see the existing institutional plurality from a gender perspective.

[7] There are many studies carried out in different part of Ethiopia that confirm a certain pattern of task division among men and women family members (Aregu *et al.* 2011; Yigremew 2000; Dejene 1994; Bishop-Sambrook 2004).

riage arrangements is critical for identifying the relative position of men and women in everyday societal life. Particularly, understanding such institutional structures helps to capture the relationship of men and women to livelihood resources.

Patrilocal marriage arrangements in all of the study *kebeles* facilitate land acquisition by men rather than women. Thus, one of the socio-economic features of such an institutional arrangement is related to women's disinheritance of property such as land. Under this arrangement, young women leave their birth village upon marriage, and have no permanent claim to the family property such as land.

Even though female heads of households under the government's current land arrangement are granted equal access to land with their male counterparts, they are not free from the negative consequences of patrilocal settlement. For instance, all the female respondents who are either divorced or widowed have already forfeited their right of inheritance upon their marriage. In addition to this, patrilocal marriage arrangements unite not only the bride and groom, but also their families as well as lineages, upon which the kinship system of the study population is based. Thus, while living far away from their family and kin, women fall under the close guardianship of their husband's relatives and kin group. This, however, comes with some costs to women in which they compromise their freedom to exercise decisions in matters that are of critical importance to their future livelihood and well-being. For instance, widows are easy targets for property claims from their deceased husbands' relatives. Various local cultural norms and perceptions, and societal pressures, such as being excluded from community organizations like *Idir* as well as *ergiman* (curse) from *shimagiles* (respected wise old men), discourage women from taking the legal course to get justice on property division following a divorce[8] (case stories and empirical findings on the impact of this institutional arrangement on control over resources are presented in chapter eight, see 8.2.1).

Value attached to the institution of marriage and family

In the study area, apart from being a sign of adulthood and the beginning of maturity for the marrying couples, marriage signals cultural solidarity and symbolizes the continuity of local tradition and societal norms. Therefore, marrying couples receive due recognition from their friends, relatives and the community which they are a part of. A woman who marries is believed to be a special source of pride and esteem for herself and her family, and the community in general.

Thus, among the Gurage people there is a strong stigma attached to being single, or being a single parent or a lone mother. FHHs generally have a low social status as compared to men who are single heads.[9] The impact of such gender-inflicted perceptions is reflected in the behaviour of female heads in the study area. During the FGDs and some of the participant observations conducted in the study *kebeles*, female heads preferred to use titles such as "*bale tidar*" or "housewives", instead of calling themselves a "widow" or a "divorcee". In explaining the desirability of living in a marriage instead of being a single mother, a widowed female head in an FGD in Beresa *kebele* said: "*I prefer to have a crippled man at home to being a female head.*"

[8] Earlier studies in other rural settings of Ethiopia also agree with my findings (Zenebework and Yared 2000; Yigremew 2005).

[9] Here it is worth mentioning that men tend to remarry immediately after a divorce or the death of their spouse, while this is rarely the case among women in the study area. Women tend to remain single for fear of stigmatization and exclusion. A woman who engages in a second marriage is considered a disgrace to her community.

The latter opinion is quite different from the more ordinary reasoning which relates having a male spouse to some expected economic benefits. Instead, the ironic expression of this respondent sends a strong message that the value attached to marriage among the study community goes beyond mere economic benefits and touches on other social and cultural dimensions of societal life. Such internalized feelings of loss of social and cultural identity, as reflected in the view of the above respondent, can have a damaging effect on women's self-confidence and self-esteem. They can also undermine women's well-being, since this loss of identity weakens their motivation and desire to exercise unrestricted decisions in various areas of their livelihood.

Local norms regarding 'the socially appropriate behaviour of housewives'

The local norms in the study area promote docility, patience, submissiveness and total obedience of women to their husbands as desirable behavioural traits within both the domestic and public spheres. Even though these norms theoretically apply to married woman living with spouses, inquiries as to whether women continue to show such behavioural traits as female heads produced interesting results. In general, whether such norms apply to female heads in the absence of a male spouse varies from individual to individual. Yet, the study found out that it is exceptional to find female heads who act pragmatically during social interactions at community gatherings. The existing reality on the ground is that the local norms and traditions continue to dictate the behaviour of female heads in many areas of life outside their homes. The poor negotiation and bargaining power of female heads during social encounters with their male counterparts is one manifestation of their submissive and obedient character. This undermines their ability to make proactive decisions in situations where they need to do so. Land-sharing arrangements such as sharecropping and mortgaging are good examples to demonstrate where unequal gender power relations prevail, as women submit to men's decisions even when this works against their interests.[10]

Harmful Traditional Practices (HTP)

Harmful Traditional Practices (HTP) in the study area are customary practices performed on girl children and women.[11] HTPs emanate from deeply entrenched discriminatory views and beliefs about the role and position of women in the Gurage society. Mostly, these practices are defended in the name of preserving the institution of traditional marriage and family in which women's relative position is simply to subdue to the will of their spouses. For instance, widely used popular old proverbs such as "*girls and baby maize are sweet when eaten fresh*" are meant to justify early marriage, abduction and other marriage-related customs in the study area. Therefore, it can be argued that HTPs are cultural practices by which violence[12] against women is maintained and perpetuated.

HTPs perpetuate dependence, poverty and vulnerability in various ways. For instance, abduction and early marriage deprives women of access to formal education, a key factor to access other livelihood resources. A development agent in Semen Shershera explained that it is hardly possible to find women who have graduated from a high school due to the large number of women who drop out of school only to be dependent on male spouses for the rest of their lives. In this regard, the survey findings also show that the majority of female-headed house-

[10] For further reading on how unequal power relation plays out in the field and its impact on control over land, see 8.2.1.

[11] According to assessments carried out in three districts of the SNNPR, including Meskan district, some of the most pervasive HTPs are rape, abduction, female genital mutilation (FGM), and polygamy (see Table 4.2).

[12] According to Article 2 of the Declaration for the Elimination of All Forms of Violence Against Women (DEVAW), violence can be physical, sexual and psychological and can occur in the family, in the community or it can be perpetrated or condoned by the State (UN 1993).

holds in the study are in the illiterate category (see Figure 6.3). HTP-inflicted illiteracy excludes women from taking advantage of new livelihood opportunities in contrast to their male counterparts. Instead, it facilitates risk-aversive behaviour of women towards new ideas and methods of production. For instance, researches in different parts of the country show a strong association between level of education and adoption of new technologies by farmers.[13]

Moreover, HTPs deprive women and girl children of their physical and psychological integrity. They rob women of their self-worth, destroy their self-confidence and their self-esteem, undermining their capability to take meaningful and appropriate decisions on matters related to their livelihood. Female heads who experienced polygamous relationships[14] for instance reported that they enjoy restricted access to the family's land and farm produce or that they enjoy no ownership rights in the family land. They cannot make decisions regarding important areas of their livelihood, such as land transaction and sale of livestock, since their social status puts them in a subordinate and inferior position.

7.2.2 Gender roles

In the patriarchal society of Ethiopia, gender-based behavioural orientations are the result of socialization processes that start in the early years of childhood (Haregewoin and Emebet 2003). In-depth interviews with local people in the study area revealed that growing up with their parents and in their community, girls and boys get different social exposure which allows them to develop distinct skills and behaviours based on what traditional norms consider "a gender-appropriate" code of conduct. For instance, boys generally tend to have early contact with activities outside the home, such as ploughing, herding and other farm-related activities. They are thought to be bold, decisive and are expected to learn and become self-reliant and major bread winners. Moreover, they exercise leadership by taking responsibility in certain family and community affairs. Thus, starting from their early childhood, they learn to face challenges and are encouraged to solve problems. Girls, on the other hand, are made to specialize in indoor activities, such as cleaning, washing clothes, fetching water, cooking, collecting fire wood and going to market. Moreover, they are expected to assist their mothers in and around the home and engage in taking care of their siblings. Generally, unlike their male counterparts, girls are trained to conform to culture, to be shy, soft, dependent, honest, obedient and submissive. The community also plays its role in these processes by disapproving norm-inappropriate behaviours and conducts.[15]

Both men and women continue to display these early behavioural characteristics in the kind of decisions and actions they make at later stages of their life. For instance, women gradually develop a feeling of economic dependency on their future husband.[16] They also show less interest and motivation towards self-development and economic reliance. On the contrary, men pursue their ambition to become economically independent by claiming a share in the family's property, such as farmland. This kind of well-established socio-cultural institutional setting facilitates the continuity of traditional gender roles in the establishment of a family.

---

[13] The findings of the study regarding gender and livelihood diversification are presented in chapter nine.

[14] There were female heads in the FGDs who went through polygamous marriages before they became lone mothers.

[15] Ostracizing and exclusion are some of the negative reinforcement mechanisms widely used in the study area as punishment.

[16] Influenced by the day-to-day realities of their mothers, friends and neighbours, girls are convinced at the outset that their position in society is pre-determined and doesn't require hard work. For instance, girls seem not to be interested in having a share in the family's property, like land and oxen. They are convinced that they will have a secure home as long as they get a husband.

For an in-depth understanding of the impact of gender roles on the livelihood of male- and female-headed households in the study area, the following section selectively highlights gender roles and division of agricultural tasks in the study area with a particular emphasis on the oxen-plough culture.

Gender-based agricultural division of tasks: the oxen-plough culture

A number of studies have been made regarding the gender-based allocation of agricultural tasks in various regions of the country. The studies show that gender-based assignment of roles varies depending on the type of farming system, cultural setting, location and wealth categories, which are different from place to place. However, regardless of variations among and within regions, most of these studies confirm that women in Ethiopia are largely responsible for nearly all domestic and reproductive tasks in addition to their role in productive and marketing activities, while men are usually household heads and primarily engage in the productive sphere (Aregu *et al.* 2011; Yigremew 2000; Dejene 1994; Bishop-Sambrook 2004).

The findings obtained in the study area do not deviate much from the findings at national level. Within the traditional family, men are generally assigned a role as household head and chief manager of the family's farmland, which is the primary source of livelihood for households in the study area. In addition to this, they are largely responsible for land operation, namely land clearing (where it is necessary), ploughing and threshing. Women are expected to take part in all agricultural activities by assisting the men. This assistance ranges from carrying farm equipment to the field to bringing farm produce to the stores. However, it should also be stressed that women undertake these activities in addition to their responsibility for managing a wide range of bulky and time-consuming domestic and reproductive tasks. With regard to the former, they engage in cleaning, fetching water, cooking and collecting fuel wood. Tending the sick and elderly, as well as domestic animals, is also primarily within the women's domain of activities. Reproductive responsibilities of giving birth and nurturing children are entirely left to women.

Against the above backdrop, in the study area, once a woman becomes a single household head, the responsibility of operating the family farmland will be left exclusively to the woman and her children. However, given the deep-seated gender-based division of labour that exclusively assigns tilling of land to men, female heads of households find it extremely difficult to take up such activities in the absence of grown-up children of their own.

The respondents who took part in the survey were asked if they engage in ploughing their land by themselves. The survey findings summarized in Table 7.1 show that around 99 percent of FHHs do not plough their farmland by themselves, while nearly 98 percent of MHHs do engage in ploughing activities.

Table 7.1 Percentage distribution of respondents who engage in ploughing

Responses	Type of Household				Total	
	FHHs		MHHs			
	№	%		%	№	%
Yes	2	1	182	97.8	184	49.5
No	184	98.9	4	2.2	188	50.5
Total	186	100	186	100	372	100

Source: Household survey, March to April 2011

In-depth interviews and FGDs were used to capture the views of female heads regarding why they refrain from engaging in ploughing. Based on the responses obtained, the reasons are categorized as culture-related and time-related factors, which are discussed below.

A. ***Culture-related factors*:** Almost all FGD participants considered ploughing culturally a man's task and, therefore, consider it inappropriate for a woman to engage in it. Yet, there were some differences among respondents with regard to the possible consequences of violating the local norm. One group said it would result in women being ostracized from the community organizations and group activities. Moreover, curse (*ergiman*) from *shimagiles* is mentioned as the other consequence that a woman might have to bear. However, others responded that it causes no harm since women have more rights now than some 20 years ago. Yet, both groups share the view that they refrain from practising ploughing for the sake of obeying and conforming to the local culture. They consider it as one way to show their obedience to the local norm and a mechanism to preserve their status as "housewives" in their community.

B. ***Time factor***: one of the most striking points raised during the FGDs is time stress as a constraining factor for undertaking agricultural tasks such as ploughing.

To quote one widowed female respondent in Dirama *kebele*:

> *"Ploughing is not a difficult task to do. After all, except ploughing, I do every household chore. I do cleaning, cooking and fetching water. And these are very very laborious and time consuming. Since we don't get water easily in our kebele, I spend hours and hours to fetch water every single day.*[17] *The day is not enough to finish my work, let alone to plough my land."*

There are at least two points worth commenting on in the widow's response. Firstly, she suggests that the reason why she does not engage in ploughing activity is more a matter of infrastructural problems than of culture. By implication, improving the accessibility of water and introducing time-saving technologies for household chores would enable women to engage in activities such as ploughing. Secondly, this reasoning that avoids the usual "culture-based argument" is an indication of existing dynamic socio-economic processes and the parallel development of women's pragmatic thinking with regard to the social environment with which they constantly interact. This is an essential pre-condition for women actors to start exercising their agency proactively, and is also an essential force in changing the existing institutional structures.[18]

[17] The effect of shortage of water on food security of households is discussed in detail in chapter eight.

[18] Proactive decision-making gradually leads to shifts in norms, beliefs and societal orders which might result in the modification or transformation of the existing gender role in relation to ploughing.

Implications of gender roles for households' livelihoods

Based on the information gathered through focus group discussions held with female and male heads of households, the consequences of gender roles on the livelihoods of FHHs and MHHs in the study area can be analysed in terms of four dimensions. These, in one way or another can be taken as a source of vulnerability to female heads as compared to their male counterparts. The four dimensions are psychological/attitudinal, decision-making, access to resources and pressure on time asset.

Psychological/attitudinal: Exclusion of women from important activities such as ploughing has negative psychological/attitudinal effects in two major ways. Firstly, it reinforces the perception that women lack the physical ability and the necessary skills for ploughing. It promotes the widespread perception of women as "weak farmers." Previous researches done by scholars in various regions of the country (Teklu 2005; Yigremew 2005; Mogues *et al.* 2009) have documented similar societal perceptions of women in different parts of the country. Secondly, women's abstinence from ploughing further strengthens women's dependence on men's labour, as they gradually internalize the belief that they are "passive actors" in the agricultural sector. The psychological and attitudinal impact of these perceptions not only undermines women's confidence; it can also have a long-term consequence by discouraging them from possessing and mobilizing other related productive resources, such as oxen, that are key inputs to ploughing. For example, female heads in the study area generally prefer to own cows and small ruminants over oxen. Such kinds of livelihood decisions might further constrain women from taking advantage of other livelihood options.[19]

Decision-making related: Women's restricted involvement in agricultural activities such as ploughing also comes with some costs with regard to women farmers' capacity for independent decision-making. This is particularly relevant to livelihood areas that have strategic importance to them. For instance, among the study population, since men are in charge of managing the family's farmland, they wield greater influence over what to plant, when to plough, how much fertilizer to use and what to sell when. This holds true for high-return crops and livestock such as oxen, which are predominantly within the men's domain. In the absence of a male spouse, FHHs still look for the approval of a male relative or a sharecropper for such decisions.[20]

Access to resources: The gender-based agricultural division of labour strengthens men's rights, access, entitlements and claims to farm land, production, agricultural inputs and market-related affairs. This is manifested in the findings of the study in respect of access to draught animals such as oxen, credit, technology and expert advice, where men are represented in higher proportions.[21] This is also supported by Mogues *et al.* (2009) who in their study confirm the association between the activity of livestock tending by boys and livestock ownership which is largely a men's domain in Ethiopia. The study also associates the low rate of cattle ownership by women with similar gender-role-inflicted causes. The livelihood of female-headed households can meet severe constraints as a result of the lack of important resources that are critical inputs for their food security.

Pressure on time asset: As a result of the traditional gender role, FHHs are required to carry a heavy responsibility, undertaking household chores in addition to their involvement in re-

[19] In the absence of resources such as oxen (a very important form of collateral for the study population), women are economically constrained to diversify their livelihoods.

[20] Discussions on patterns of decision-making are presented in chapter eight.

[21] Detailed discussions on gender and access to resources in relation to food security are presented in chapter eight.

productive and productive activities. Their relatively diverse roles in the society as compared to their male counterparts hardly leaves them time to participate in other community activities and trainings. In this regard, as survey findings in the study area clearly indicate, most female heads have less education than their male counter parts.[22] Women farmers in general are less likely to occupy leadership positions in community affairs and, thus, have weak bargaining power to influence decisions in areas of their livelihood that matter most to them. Such circumstances aggravate their vulnerability to poverty and food insecurity as compared to their male counterparts.

7.2.3 Customary judiciary system: the *shimagiles* as intermediaries

Customary and religious laws play a prominent role in the administration of justice and settlement of disputes in the study area. The third parties who act as an intermediary between the disputants are known as *shimagiles*. The *shimagiles* are a group of men with traditional authority to protect the local tradition and culture, mediate conflicts between people, and ensure peace and harmony in their villages. They are the most respected and widely accepted people in the society. They are usually elderly men, including chiefs of the clan or the community, religious leaders, or local administrators who have official positions. They are expected to be wise, fluent in speech, and well experienced in settling disputes. When it comes to their nomination, *shimagiles* are nominated by the local community itself to serve as an intermediary in certain cases. Customary dispute settlements in the study area have a wide scope in settling family, succession and property disputes. Very often, crimes involving theft and physical or psychological injury such as rape, abduction, polygamy disputes between spouses over properties such as land and other resources are brought to the *shimagiles*. In general, passing over the *shimagiles* and taking cases of family dispute before a formal court of law is considered shameful for the spouses.

Even though there is no formal procedure, most dispute settlement processes start with an investigation into the nature of the dispute and the behaviour of the disputants by holding a series of talks with both parties. In the conciliation process, the *shimagiles* persuade the parties to come to a common consensus. The Amharic proverb *"shimagile washto yasitarikal"*, literally "the *shimagiles* can lie to get the parties reconciled", means that *shimagiles* use various techniques to smooth out difficulties and appease the conflict. Usually, a disputant who is declared to be the wrongdoer will be required to compensate the victim either in kind or in cash. If the wrongdoer is not capable of doing so, he is helped by his clan members and relatives. It is considered an insult to not respond to the verdict pronounced by the *shimagiles*. Moreover, it is very uncommon to resist the authority of the *shimagiles* because of the social norms and grave consequences for the disputants.

The fact that violence against women is an overriding issue that women have to deal with in their everyday social life makes the role of the customary judiciary system in the study area very important. The customary dispute settlement mechanism is a useful device for creating peace and harmony, yet the whole proceeding is not free from problems. In fact, the supreme power of the *shimagiles* in the social and political life of the study community is often mentioned by key informant gender experts as one of the most challenging aspect in the empowerment process.

The following case story of a female head who is a victim of polygamy in Dobo Toto *kebele* reveals the kind of subjugation some women have to suffer while their cases are being handled by traditional intermediaries.

---

[22] See Figure 6.3 on educational background of respondents

Case study

She is a woman in her late 40s in Dobo Tuto kebele. Her husband took on a second wife who moved into the same house she and her husband had lived in for many years. The very first day he brought the new wife home, he removed his first wife's blanket and slept with the new wife. When she told her women neighbours this story, they empathized with her by putting butter on her head (a customary practice meant to bring emotional comfort to the victim). She tried to remain patient for some time until one day the conflict escalated when he sold oxen and a sack of teff without her knowledge and consent. She went to the kebele administration office to present her case and to seek legal justice. During the process, her case came to the ears of the shimagiles and a woman Development Agent who by chance was there for a meeting. The DA advised her to follow up her case through the formal court by directing her to the Women's Affairs Office in Meskan district. However, the shimagiles interfered and tried to calm down the victim by teasing her, saying "Is it not a tradition? How can you be afraid of a mistress?" They put pressure on the victim not to take the legal course, arguing that it was a family issue and should be handled according to the local tradition. The shimagiles tried to settle the dispute, as a result of which the husband was requested to take his new wife away from the village where his first wife lived. Moreover, they ordered him not to thresh the family's teff for a specific period of time, giving the kebele administration the mandate to protect further abuses with regard to crop yield. Nevertheless, her husband had done a lot of damage to the family property by selling livestock and removing all the cattle to other kebeles. And he did not divorce his second wife, even though polygamy is banned under the government law. The victim is now a female head and her livelihood is very insecure.

As shown by the case story above, it is a basic understanding that the male-dominated traditional court in the study area does not allow poor women to challenge the sanctions imposed by the court. In addition, certain important points are worth mentioning here. Firstly, the fact that neighbours showed verbal sympathy rather than giving practical support can be interpreted either as society's fear of criticism from the *shimagiles* or it might be taken as mere enthusiasm for culture and tradition. In either case, it is important to underline that neighbours and community members can also contribute to the escalation of the crime by remaining neutral or by remaining passive after it is committed. Secondly, even though the formal legal course is available to women under the present law, poor women are less likely to embark on the long and costly process due to a lack of knowhow about legal procedures and limited financial capacity. In other cases, physical distance from the offices for women's affairs or police stations, which are only available in the district town, prevents women from taking legal recourse. Yet, other factors such as a tendency to avoid punishments such as exclusion from *Idir* and other community activities can keep women victims passive at times when violence occurs. All these factors give the perpetrators of a crime a good opportunity to hijack the process of justice and continue further abuse. Women easily lose their access to, and control over, important livelihood assets, which eventually weakens their capacity to respond to external livelihood shocks and stresses, exacerbating their vulnerability.

Generally, some of the most frequently mentioned problems associated with the justice system of the traditional court in the study area include corruption and laxness of the legal system. A key informant interview with a female Development Agent (DA) who has long experience of working in Dobo Tuto *kebele* revealed that invariably, all matters handled by *shimagiles* end up by perpetuating violence, encouraging it further or directing the woman to live in violence. Most of the injustices occur when the well-being of the victim is compromised for the sake of preserving the institutional status quo according to the local norm and culture.

7.3 Local institutions in transformation: prospects and future challenges

The social and economic transformations happening elsewhere have caused changes in the social and economic structures and processes in the study area. In relation to this, it is worth noting that the pattern of gender-based farm- and off-farm-related tasks within the traditional household institution is gradually changing, regardless of the importance given to the culturally ideal position of men and women in their community. This is manifested in the study *kebeles* where women and men are found switching and shifting their traditional roles, especially at times of livelihood crisis.[23]

According to the data collected from the district land administration office shown in table 7.2 there are around 14 types of land-related conflict coming from women, mainly divorcees and widows, while only 3 types are identified in the case of men. This shows the diversity of the legal issues raised by women as compared to men.

Table 7.2 Dispute cases reported to Meskan District Land Administration Office by women and men (2005-2010)

№	Cases filed by women	Cases filed by men
1	Battering	Claim in respect of land possession
2	Loss of land	Conflict between siblings
3	Forbidden to cultivate on family land	Complaint about land measurement
4	Spousal conflict over shared property	
5	Property conflict in polygamous relationship	
6	Conflict over division of property	
7	Conflict over property sharing between siblings	
8	Conflict over inheritance	
9	Claim in respect of land ownership	
10	False witness	
11	Threats against one's land possession	
12	Spousal rights	
13	Forgery of land certificates	
14	Conflict over grazing land	

Source: Own compilation from Meskan District Land Administration Office, data obtained on 11/3/2011

Due to heavy restrictions imposed by law and the awareness-creation campaigns over the last decade, the popularity of some of the customary practices is fading away in the study area. Women are resisting some of the HTPs and gender based violence and are seeking justice through legal recourse now more than ever before. For instance, more and more cases over rights, polygamy and polygamy-related property claims are being brought to the formal court, rural land administration offices and the police, instead of to the *Shimagiles*. Figure 7.1 below shows the number of cases brought to the court by female and male household heads. At aggregate level, the figure shows a decreasing trend in the study *kebeles*. Yet, disaggregated

[23] Some of the findings in the study area that prove the dynamic nature of gender roles are discussed in subsequent chapters.

along gender line, the number of court cases filed by FHHs shows an increasing trend vis-à-vis MHHs.

Figure 7.1 A gender disaggregated trend of cases reported to Land Administration Office (2005-2010)

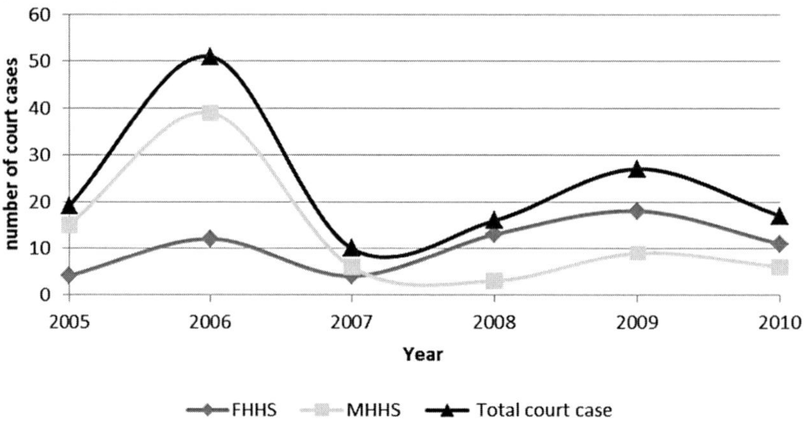

Source: Own compilation from Meskan district Land Administration, data obtained on 11/3/2011

Some of the HTPs have also gone through noticeable changes. A key informant interview in this regard revealed that female genital mutilation, abduction and early marriage have gone through some changes. Some of these are being abandoned while others have continued in a different form. A good example in this case is the gradual reduction of cases of abduction, which have given way to a new practice called *werchegn*, which can be roughly translated as "let's go." In this kind of marriage arrangement, a man and a woman agree to run together away from their families instead of a man secretly abducting a girl. This happens especially at social gatherings where young people get together for chewing *chat* and dancing. *Werechegn* has to some extent reduced the costs of feasting associated with traditional weddings. However, this kind of marriage arrangement has increased the number of women who drop out of school. And very often, these unions fail, with women emerging as the victims without any reasonable legal basis to claim property rights when the break-up occurs. As a result, it is leading to single parenthood where women eventually become female heads.[24]

The incidence of polygamy has also reduced in the study area for certain reasons. An elderly man in his 80s from Dirama *kebele* explained:

"I had four wives, two of them have died. I have 17 children from all of them. I was rich enough to hire a farmer to plough my land. Now everything has changed. Now, I have only a small plot of land and my two wives have claimed their share as well. It was because the time was good that I had four wives. Who can marry more than one, these days? Some are praying to God for one, let alone three or four wives. You know why? The land is just 2 timad[25]!! (Laugh...) And again, the woman will have a share in it!"

First of all, most men are economically constrained and can no longer afford such marriage arrangements, unlike in earlier days. Secondly, as a result of improved laws and legislations,

[24] Investigating the socio-economic setting and the impact of this new practice on the lives of women would be a topic for further research.

[25] The equivalent of 0.5 ha.

the number of women who demand their rights in respect of land and other household property has increased. Both FGD discussions and in-depth interviews revealed that men go for polygamous marriage for reasons like having more children. Yet, despite their desire, it is now becoming costly for them, as women become more aware of their rights.

7.4 Interim summary

Presented in two main sub-sections, this chapter captures the existing vulnerability context of women in relation to their male counterparts in the study area. In doing so, it started with the presumption that in the face of institutional pluralism, informal institutions largely shape the vulnerability context of female-headed households. It first shows how farm households perceive social, economic and cultural shifts and changes in their external environment against certain standardized norms, values and customary practices with which they most identify themselves. The key changes include the unpredictability and variability of rainfall; disruption of the local agricultural calendar; growing dependence on market, stress in respect of labour and time allocation at household level and disintegration of mutual and reciprocal social support systems.

The study concludes that the impacts of these changes are felt differently along gender lines, causing worse vulnerability conditions for female-headed households than for male-headed households. The situation and underlying causes are discussed within the context of institutional plurality. The three institutional areas considered are traditional marriage and family, gender roles, and the customary judiciary system as the most relevant institutional mechanisms through which the vulnerability context of female-headed households in the study *kebeles* can be understood.

Assessment of the implications of these institutions on local livelihoods reveals important evidence that illuminates the greater degree of vulnerability of Gurage women as compared to their male counterparts. The findings show that the position of women in the traditional marriage and family institutions is backed by a number of harmful traditional practices and customary arrangements that legitimize violence against women as an acceptable way of life. The findings also show that despite the contribution of gender roles in maintaining the traditional family and household structures, the gender-based division of tasks reinforces the subordination and exclusion of women from important decision-making processes. For instance, women bear the heaviest burden of tasks within and outside the domestic sphere, trapped within their double role, while prevented by time constraints from acting in the public sphere. In this case, FHHs as sole providers for their household, are severely affected. When it comes to HTPs, customary practices such as early marriage and abduction that are widely used in the study *kebeles*, have denied women the opportunity to exercise their right to education, a key factor in accessing all other economic, social and political resources for building a secure livelihood. Moreover, violence perpetuated through HTPs erodes women's confidence and self-esteem, creating more economic dependence on men, and exacerbating their level of insecurity when confronted with livelihood shocks and stresses.

Most of the institutional elements in the study area promote unequal gender power relations that are, however, dynamic by nature. In some cases, these unequal gender power relations are undergoing modification, while in other cases they are being reproduced through the actions of both men and women. The continuing external socio-economic and policy dynamics are also facilitating this process. For instance, HTPs such as the dowry system, early marriage, abduction and polygamy have gone through certain transformations over time. However, there is only a slow process of change in the areas of gender roles and the customary judiciary system. In these areas, the existing unequal gender power relations continue to guide social interactions in local livelihoods. Based on the findings, it can be argued that social life in the study area shows a moderate level replica of Giddens' structuration process. This is substantiated by the fact that the actions of men and women in the study area continue to influence the reproduction and modification of the local informal institutions, while these institutions continue to shape the decision-making and actions of rural men and women.

To sum up, due to the patriarchal system which dominates all local informal institutions, women in the study area face life-long discrimination in all walks of life, leading to their marginalization and social exclusion in their day-to-day livelihood routines. As a result, they find it difficult to effectively respond to the external livelihood shocks and stresses that are discussed in the first sub-section of this chapter. This implies that the constraints to women's empowerment are not only economical and physical, as many of the government programmes emphasize, but rather socio-cultural and attitudinal. And this is deeply entrenched in the various informal institutional forces discussed in this chapter. Thus, from the analysis made so far, it follows that the existing evidences in the study area support the hypothesis stated earlier.

8 Determinants of household food insecurity

The study hypothesized that with less access to resources and poor decision making, FHHs are more food insecure than their male counterparts. In order to respond to this presumption, this chapter is divided into three sub-sections. In the first sub-section, the households' pattern of food gap is assessed through a self-reporting method by which individual respondents were directly asked whether they can meet food demands all year round from their own production. The second sub-section presents a comparative assessment of access to livelihood resources in female- and male-headed households. The chapter winds up by presenting the study findings regarding the interaction between decision-making and food security in the study area.

8.1 Households' perception of their food security situation

Food security has been defined in various ways. In the context of this study, the author finds it useful to adopt the definition used in Degefa (2005, 10) in which he describes a food secure household as the one who is able to meet food requirements and other basic needs, either through own production or through market ventures.

For the purpose of this study, a question regarding the average period of coverage of food consumption through the household's own production was forwarded to the survey respondents to assess the extent of the contribution of their own agricultural production in meeting their household's food demand within one year. Accordingly, households were asked to roughly give the number of months they are able to cover their household food needs under good rainfall conditions.[1] Figure 8.1 shows the findings in each of the study *kebeles*.

Figure 8.1 Period of food consumption coverage under good rainfall conditions by *kebeles*

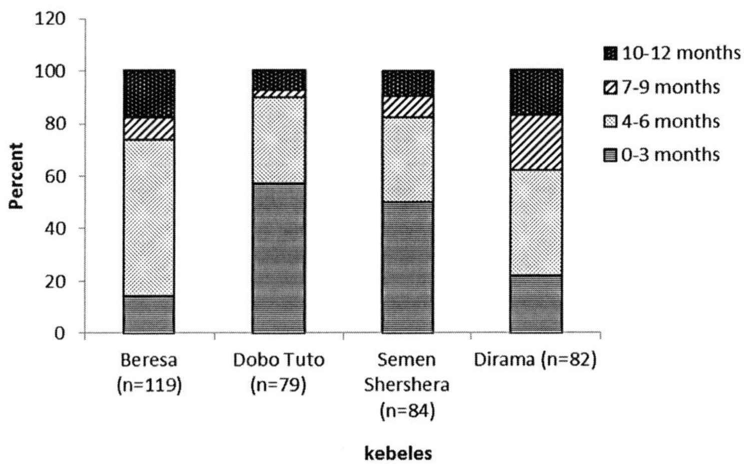

Source: Household survey, March to April 2011

The survey result shows that the extent of consumption coverage varies among *kebeles*, ranging from 0 to 3 months to 10 to 12 months. A comparison of the figures across the four study

[1] In the study area the condition of the *belg* rains determine the severity of the hunger season, since they are important for producing crops that assist farmers to overcome the "hunger season" (June - August) and make the transition to the September and October harvest season. The *belg* rains are highly variable in the study area (see Figure 4.3).

kebeles shows that Dobo Tuto and Semen Shershera perform badly, with 57 percent and 50 percent of respondents being able to cover their consumption needs for less than three months. The situation is better in Beresa and Dirama *kebeles*, where a larger number of respondents, nearly 60 percent and 40 percent respectively, said that their own production covers their consumption for nearly 4-6 months in one year. As a general observation, the survey result suggests that under normal rainfall conditions, agricultural production does not meet the consumption needs of the majority of households for at least half of the agricultural year. As a result, serious food gaps characterize all of the study *kebeles*, despite slight differences among them.

The responses obtained from FGDs with both female and male household heads confirmed the seriousness of the problem of the food gap even under good rainfall conditions. This is regardless of the gender of the household head. The survey result regarding the gender dimension of food gaps in the study area is presented in Figure 8.2.

Figure 8.2 Period of food consumption coverage under normal rainfall conditions by type of household

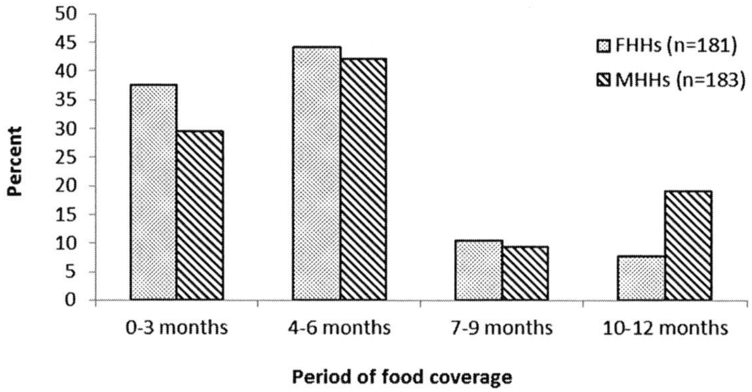

Source: Household survey, March to April 2011

Disaggregated along gender lines, around 38 percent of FHHs as compared to 30 percent of MHHs reported that their own production covers only up to three months of their family's consumption. Moreover, the survey result shows a significant gender gap in the category of 10-12 months coverage, in which the percentage proportion of MHHs is more than double that of FHHs. Pearson's chi-square result (x2 = 10. 76) at 5 percent level also proves that there is a significant difference in terms of food gap between female- and male-headed households under normal rainfall condition.

In terms of the local agricultural calendar in the study *kebeles*, households face a food gap starting from the main rainy season in June until the start of the green maize harvest around the end of September. This period signals the main "hunger season" in all of the study *kebeles*. Food purchases soar during these months. This is also a period in which livestock prices drop, as many households sell animals in order to obtain cash to purchase food.

To gather further information on households' food security situation in the study area, the views of respondents regarding their food security situation was gathered using a self-report checklist. To reduce subjectivity, the meaning of food secure and food insecure households was explained to the respondents beforehand.[2] Then, respondent households were able to identify their household food security situation under the following categories: (a) households that live with food shortage crises and subsequent hunger throughout the year/ acute food shortage; (b) households that face food shortage problems only when they are hard hit by disasters and shocks, or are temporarily or seasonally food insecure, and (3) those who consider themselves food secure throughout the year.

The findings showed that a large proportion (71 percent) of the households in the study area felt seasonally food insecure. Nearly 18 percent felt that they are food secure throughout the year, and the remaining 11 percent felt that they are food insecure throughout the year. Yet, it should be underlined that households who reported food security throughout the year mostly depend on food purchases to fill food gaps that are created due to food shortage from their own farm production. A comparison between the four *kebeles* in this category shows that Dirama is far better off than the rest of the *kebeles* with nearly 19 percent all year round food secure households. On the contrary, the situation in Beresa is the worst with the least number of households in this category. Generally, the survey result indicates that an overwhelming proportion of households in the study areas live under seasonal or acute conditions of food shortage.

Looking at the survey findings in female- and male-headed households, as presented in figure 8.3, MHHs are seen to be far better off than FHHs in the category of those who reported food security throughout the year. There is only a slight difference between female- and male-headed households in terms of food insecurity throughout the year. Yet, a greater proportion of FHHs (nearly 77 percent) identified themselves as seasonally food insecure than male-headed households (nearly 65 percent). Moreover, as expected the result of Pearson's chi-square test ($x^2 = 9.99$) at 5 percent level shows there is a significant difference between FFHs and MHHs in relation to household food security status.

Figure 8.3 Household food security statuses by type of household

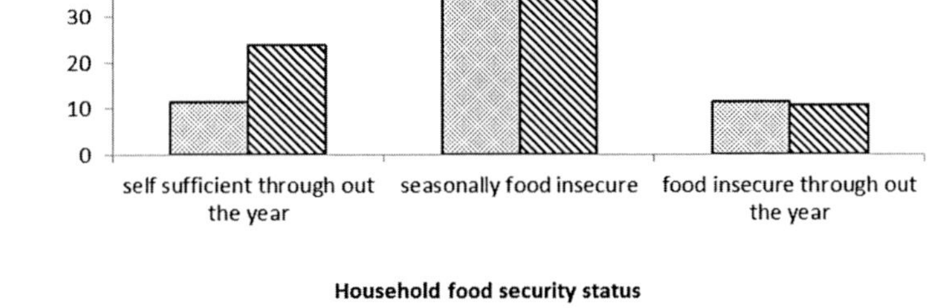

Source: Household survey, March to April 2011

[2] A food secure household is defined as one which is able to meet its food and other basic needs all year round using its own production or through purchase from the market by deploying its own assets.

From the above finding, food insecurity is a common problem for both FHHs and MHHs. Yet, in terms of the degree of food insecurity, FHHs are more affected by seasonal food insecurity than their male counterparts. The institutional barriers discussed in the previous chapter have a direct or indirect impact on the access of women to key livelihood resources, leading to such a gender gap. The subsequent sections focus on these determinant factors, using illuminating empirical examples.

8.2 Determinants of household food insecurity: access to resources

Many of the food crises observed in the study sites have been aggravated by man-made and natural calamities. These environmental hazards[3] have cost people's lives, damaged crops and livestock, caused asset dispossession and led to the displacement of people. However, both chronic and transitory food insecurity problems as experienced by most farm households have been due to the gaps created between food production and consumption, weak coping strategies against food shortage, and lack of regular and sufficient income to enable purchase of food on the market. To support this with empirical evidence, households who participated in the survey were asked to reflect on the situation with regard to crop yield by comparing the year 2010 with the previous two years. Table 8.1 presents the responses of FHHs and MHHs.

Table 8.1 Respondents' response concerning the 2010 crop yield by type of household

The 2011 crop yield compared to the previous two years	**Type of Household**				**Total**	
	FHHs		**MHHs**			
	№	%	№	%	№	%
Increased	23	12.9	30	16.5	53	14.7
Remained constant	43	24.2	21	11.5	64	17.8
Decreased	112	62.9	131	72	243	67.5
Total	178	100	182	100	360	100

Source: Household survey, March to April 2011

The majority of both FHHs (63 percent) and MHHs (72 percent) reported that their crop yield during the year 2010 was lower than in the previous two years. The figure for MHHs is slightly greater than for FHHs. The percentage of FHHs who observed no change in production is nearly twice that of MHHs. Even though it is difficult to guess the connotations that might be attached to this kind of response, it might be associated with failure to keep regular yield records.[4] In general, it is important to underline that a large proportion of households who reported production decrease means more pressure on food reserves and more demand for other strategies to fill food gaps.

A range of factors are identified by respondents as reasons for low production in the year 2010. These include crop damage, input-related constraints including quantity and quality, irregularity of rainfall, shortage of land and poor soil fertility. Moreover, through FGD discussions and key informant interviews held with female- and male-headed households the study also shows that infrastructural constraints, such as lack of proper connecting roads to markets, and lack of livelihood assets and skills to diversify livelihoods, aggravate the food

[3] The subsequent section deals with some of the hazards that continue to challenge households' food security and capacity to cope.

[4] In the case of FHHs, the lack of education and social factors, such as the non-participation of women in ploughing, might discourage them from keeping track of the crop yield on their farm land.

gap by limiting access to food through purchase. To simplify understanding, the responses obtained are divided into two categories. These are access to livelihood resources and hazard-related factors. This section focuses particularly on the former from a gender perspective.

Access to resources

There are sufficient studies showing the relationship between assets and vulnerability and thus food insecurity (Moser 1998; Getachew 1995; Degefa 2005; Devereux 2000). The following section presents the findings of the study on access to resources as a key determinant factor of household food security. Likewise, this section discusses the impacts of certain institutional mechanisms on access to resources and the food security of female- and male-headed households. In doing so, the special role of institutions in household food security is further substantiated using empirical evidence. Given that resources are also the medium through which power in the form of decision-making and agency is exercised (Giddens 1979), it will be shown how the interaction between resources and decision-making plays out to affect decisions in pursuing livelihood strategies, including coping strategies in respect of food security.

8.2.1 Access to and control over land

A. Land ownership, mode of access and land-sharing arrangements

The survey finding in the study *kebeles* reveals that almost all of the study respondents in the study area own some form of land. There is also hardly any gender gap in terms of land ownership per se, for nearly 97 percent of FHHs own some form of land as against 99 percent of MHHs. Most of the land a household owned at the time of data collection was allocated by the state during the land re-distribution program. Yet, both the state and family inheritance were the dominant sources of access to land for a large proportion of FHHs and MHHs before the current administration of rural land distribution. Following the classification emphasized in de Janvry and Sadoulet (2001), inheritance through the family takes both the inter-generational transfers in the form of inheritance after a parent's death or inter-vivos transfers and through intra-household allocation of plots to specific members.

In the study area, the custom of patrilocality hardly allows women to bring land into the marriage, unlike their male counterparts. This coincides with earlier findings of Yared (1997). Also relevant here is the spousal allocation of land upon marriage dissolution. In the case of divorce, land is split between the husband and wife. In the case of the death of their husband, women are either able to keep all the family's land, or it is split among the deceased spouse's children if there is no claim from the deceased's family and relatives.

The other way of obtaining land, in addition to the state and the family, is through transactions such as renting, loaning, purchase, mortgaging and non-market-mediated means of land acquisition such as gifts. Discussion on forms of land transaction in the study area revealed that an informal lease-based contractual agreement (mortgaging) for a period of 4 or 5 years, or a system of shared tenure arrangement known as sharecropping[5] are widely used in the study area. The maximum size of land that households in the study area are allowed to own through various land transaction channels other than the official means and inheritance is 10 *timad* [2.5 hectare (ha)].

[5] A system where usually male farmers enter into agreements to plough someone's land for shared and mutual benefits.

In the study area, informal rental land markets primarily take a form of share cropping which is common among FHHs.[6] There are generally two approaches that are followed in the process of sharecropping in the study area. The first one, which is also the most widely used approach, takes place when a woman enters into a contractual agreement with a farmer for a specific agricultural production season in which he contributes his own oxen and labour in exchange for half of the produce at the end of the production season. The second type of arrangement, known as *"gebere magbat"*, is becoming less popular. In this kind of arrangement, a woman hosts and provides accommodations for a farmer during a specific production year in which he operates on her farm land in exchange for half of the production. In this case, depending on their agreement, the farmer might take up responsibilities such as cattle herding. Since such kinds of land-sharing arrangements involve bargaining and negotiations between the sharecropper and the owner of the land, contractual agreements are mostly made in the presence of *shimagiles* who can later serve as mediators if conflicts arise. The type of decisions made as a result of the negotiations might vary for reasons which this study did not capture in full. For example, a key informant from Semen Shershera *kebele* revealed that she prefers not to share the costs of fertilizer just to avoid claims of hay. Yet, reports from other *kebeles* such as Dobo Tuto revealed that agreements on sharecropping are gradually being transformed to include sharing of hay by the sharecropper due to the falling trend in crop yield.

Most female FGD respondents associate sharecropping with low crop yield for various reasons. These include problems related to farm management and follow-up, starting from the time of land preparation up to harvest. Delays in timing in relation to ploughing and application of agricultural inputs such as fertilizer are mentioned as the most common problems. The impact of low production on the food security of female-headed households consists of both direct and indirect consequences. In the case of the former, it affects food security by directly influencing the availability of food at household level. In the case of the latter, it indirectly reinforces perceptions that weaken women's bargaining and negotiating power during land contractual deals, due to preconceived views concerning the low productivity of land owned by FHHs. It also undermines their creditworthiness in the eyes of both formal and informal creditors.

B. Land size

In Ethiopia, land legislation clearly gives men and women farmers equal rights to agricultural land. Yet, women farmers are generally known for minuscule holdings. The plot of land under their ownership is usually marginal and infertile, so that sometimes it is equivalent to landlessness. Several studies conducted in different regions of Ethiopia have shown that on average FHHs have smaller plots of land as compared to MHHs. For instance, a study by Sara and Muluneh (2013) in Cheleleka Bobe and Gare Kora *kebeles*, by Etenesh (1999) in Ada district, central Ethiopia, and the work of Dejene (1994) in East and West Showa all prove the above assertion. Likewise, the findings in the study area show that nearly three quarters of the female-headed respondents possess below 0.5 ha of land, whereas the figure stands at nearly 55 percent in MHHs (figure 8.4). Even though there are generally few households that own above 1 ha of land, the proportion of FHHs (nearly 4 percent) in this category is also lower than MHHs (7 percent). The computed value of average land size for FHHs and MHHs is 0.5 ha (2.1 *timad*) and 0.7 ha (2.7 *timad*) with a standard deviation of 0.287 and 0.403 respectively.[7] Generally, the findings prove that on average FHHs own smaller plots of land than their

[6] One of the limitations this study met is the difficulty of getting quantitative information to show the gender gap in the modes of access outside of the state channel.

[7] The stand deviation is computed using hectare as the unit of measurement. 1 hectare is 4 *timads*.

male counterparts. Moreover even though the average values for both categories seem very close to each other, the higher values of standard deviation in MHHs as compared to FHHs shows a greater distance of the figures for MHHs from the average value. In the case of FHHs, the values are at a lesser distance from the average value.

Figure 8.4 Land size by type of household

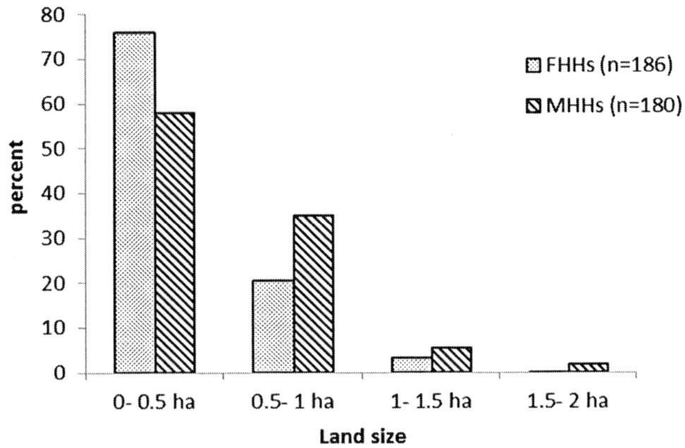

Source: Household survey, March to April 2011

Group discussions with female heads with no land or very small plots of land revealed that patriarchal systems work against women with regard to accessing large plots of agricultural land. As shown above, one of the biases is reflected in the existing gender gap in terms of land size in the study area. Factors such as lack of inheritance, gender roles and patriarchal misconceptions, combined with their limited access to other critical resources[8] and services such as oxen and credits, work against their ability to acquire more land through state and non-state channels. For example, during a FGD discussion a female respondent told of her experience with land claims. She was unable to get land from her *kebele* land administration office because of her absence during land allocation. She also explained that she has found it impossible to acquire land through the market due to her poor economic status. Generally, qualitative information on land acquisition by FHHs in the study area demonstrates the limited alternative options for economically poor women. Yet, it should be underlined that access to land through market channels might be an essential component to be considered for resource-poor FHHs, since land acquisition might facilitate access to other productive resources, such as credit, an essential livelihood input for households' food security.

Small plots of land limit crop yield and impose a unique challenge upon FHHs. Due to its small size and the meagre benefits associated with it, land owned by FHHs is less attractive to sharecroppers and is undervalued by them in negotiations. The FGD participants from Semen Shershera revealed that the problem of obtaining a satisfactory yield is exacerbated by the poor fertility of the soil and the marginal land which most FHHs own. These factors lead to a higher risk of food gaps.

C. Control over land

To what extent are women farmers able to protect and maintain their access to agricultural land? In Zenebework and Yared (2000) it is strongly emphasized that despite the immense

[8] Oxen and credits are a pre-requisite for acquiring land through state and non-state channels.

importance of having access, this is not sufficient without effective control. Government grant of land use and transfer rights is not sufficient if women are not able to benefit from them in practice. In this regard, it is important to emphasize the existence of certain institutional barriers in the study area that work against women enjoying full control over their land, even when they possess it. Among these are the existing formal legal institutional rights, which usually come with heavy costs to poor households, such as female-headed households. For instance, constant cultivation and residency requirements, which many FHHs cannot adequately fulfil, can be mentioned as a primary challenge. Under conditions where such criteria are not met, women farmers become vulnerable to losing their land.[9] Weak institutional mechanisms and processes to properly enforce women's land rights, and institutional pluralism, constrain women farmers from enjoying control over their land. The study seeks to support this statement with the following case story of a widow living in Beresa *kebele*, who was forced to give up her land to a farmer to whom she contracted it.

> *"I contracted my 1 ½ timad land to a man because I badly needed money. After the man had ploughed my land for some time he finally claimed that the land belonged to him. I took my case first to the district and then to the zone. I travelled 3 times to follow up my case. It cost me a total of 3800 birr (148€). My case was going well until one day when the man fabricated a fake document and put pressure on officials at local, district and zonal[10] level to pass a verdict in his favour. Then, I was forced to lose my plot of land to this offender. At that moment, I had no choice but to give up because I did not have money and relatives to support me during the court procedures."*

This story is a clear manifestation of the kind of resource-related institutional barriers female heads face in their day-to-day social interaction with their male counterparts. The level of agency of men and women with conflicting interests shown in the case story above shows the disequilibrium in gender power relations. In other words, the effect of unequal gender power relations is seen in negative outcomes such as the loss of land, as experienced by the woman in the story. A complex set of factors such as lack of education, finance and social capital, and gender-biased norms, together affect the bargaining power and the agency of the victim in the story.

In analysing the causes of loss of control among divorced women, Fafchamps and Quisumbing (2000) in their study carried out in Tigray, Amhara, Oromiya, and Southern Ethiopia revealed that women find themselves in a lower economic position at the beginning of household formation, which weakens their claims at the time of marriage dissolution. They continue to face similar issues for the rest of their lives as heads of households. Another case story of a widow in the study area, presented below, strengthens this assertion.

Case study

"*My land was taken away by my deceased husband's brother. He claimed that the land belonged to his father and therefore that gives him the legitimacy to protect his fathers' property. Hence, I should return it to his family. I reported the case to the police and I could not go further because I lost patience to follow up the long process that it takes to get justice. But those who have a certificate think that they are secure in the ownership of their land.*"

A widow in Dirama kebele

[9] According to the 2005 land proclamation Article 10, access to land requires permanent physical residence, ability to farm continuously and to meet administrative obligations.

[10] An administration structure above districts in Ethiopia

Customary laws in the study area that generally discourage inheritance by daughters also reinforce the idea that divorced women or widowed women without children are transients and not permanent dwellers in their husband's village. As a result, certain gaps might appear in the process of land allocation and land certification might get delayed. In the case story above, the claim by the deceased husbands' brother in respect of the family's plot of land should be understood against these institutional gaps.

8.2.2 Oxen possession

In the study area, oxen possession is an important wealth criterion with multiple uses in households' livelihoods. Key informant experts at *kebele* level describe cattle production in all of the study *kebeles* as extremely low. According to the survey result presented in Figure 8.5, the average oxen ownership, 0.9 oxen per household, clearly demonstrates the dearth of oxen possession among the study population. Moreover, the gender disaggregated survey result shows marked gender differentials in relation to oxen possession. Nearly 52 percent of FHHs against 40 percent of MHHs do not possess oxen. On average FHHs also own fewer oxen than MHHs, the figures being 0.6 and 1 ox respectively. Possessing more than two oxen is generally rare both in FHHs and MHHs, even though the figures are better in MHHs as compared to their female counterparts.

Various factors are responsible for low oxen ownership in the study area. For instance, the increasing cost of fodder, which is usually purchased from the market, and shortage of water, are common problems that have constrained oxen possession in FHHs and MHHs. Yet, additional factors are also responsible for the limited ownership of oxen by FHHS. First, women tend to bring fewer assets into their marriage and they tend to lose them if the marriage is dissolved. Moreover, their smaller landholdings when they become female heads and their lack of involvement in ploughing activities further discourage oxen ownership in FHHs.

Figure 8.5 Oxen possession by type of household

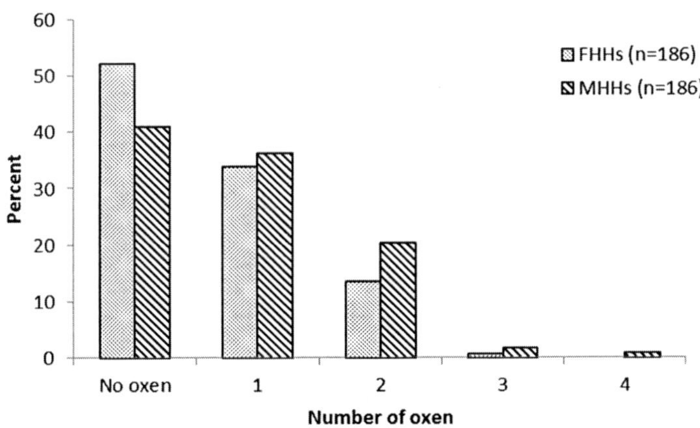

Source: Household survey, March to April 2011

Oxen being one of the most critical assets among the farming communities, lack of ownership, coupled with gender-biased perceptions, prevent women from becoming competent farmers. This puts a restraint on their ability to exploit available rural services. For instance, lack of oxen[11] restricts their access to credit. This in turn limits their financial capacity to engage in other non-agricultural ventures. FHHs with no oxen receive less attention during land allocations through official channels. In general, without oxen women are unable to exercise their power in decision-making which leads to poor livelihood outcomes, such as food insecurity and weak coping capacity. Various studies have shown similar scenarios and gender biases in different regions of the country (Frank 1999; Dejene 1994; Dessalegn 1994).

8.2.3 Labour and time constraint

> *"Even as we speak of women being empowered, they are not in any real sense. Women are still extremely burdened with domestic, productive and reproductive activities."*
>
> *A female DA in Dobo Tuto*

The above opinion sheds light on the tedious and time-consuming tasks women are responsible for in their day-to-day life. It also underscores the importance of labour and time in the livelihood of women farmers, particularly in terms of empowerment. It can also be inferred that empowerment might fail to meet its objective if it does not curb the size of the work load through which women in the study area are trapped. Against this backdrop, the following section attempts to present the findings of the study on availability of labour and time by comparing qualitative data obtained from FHHs and MHHs in the study area.

Availability of labour

Pressure on labour availability is reported by both FHHs and MHHs in all of the study *kebeles*, yet what should be underlined here is that, unlike MHHs where there is a sharing of tasks among family members, FHHs seriously suffer from the absence of such labour mobilization at household level. First, FHHs generally tend to have smaller households than their male counterparts (see Figure 6.6). Second, FHHs are constrained by a shortage of male labour. As a result, they generally engage in informal labour-sharing arrangements (see Figure 9.4)

Appropriation of time

Exploring how people allocate their time is a useful approach to assessing the quality of life people enjoy and their social and economic well-being (Crosbie 2006). Women in the study area generally spend long hours working both in agricultural and non-agricultural activities. To investigate whether there is variation depending on the type of household, it will be useful to look at how men and women divide their time between agricultural and non-agricultural tasks, leisure, travel and social and cultural interaction. However, it was found difficult to develop an activity diary that uniformly applies to female head and male head respondents. This is primarily because both female- and male-headed households depend on diversified activity portfolios which are hard to summarize. Instead, it was decided to focus on certain daily routine activities commonly shared by female- and male-headed households, in order to show the gender gap in relation to time stress and its impact on household food security in FHHs and MHHs.

Accordingly, fetching water and going to market are the two most important time-consuming activities shared by female- and male-headed households in all the study *kebeles*. Women and children are solely responsible for collecting water, while visiting markets and undertaking petty trading is primarily a women's domain. Nearly 3-4 hours are spent daily on fetching

[11] Oxen are the most trusted form of collateral for credits.

water, depending on the distance between homestead and water point. The duration increases depending on the length of the queue at the water point. Moreover, walking to and from the market takes an average of 4 hours. The time and energy spent collecting water and going to market by family members both in FHHs and MHHs is immense. The work pressure on women in general, and on female household heads in particular, cannot be understated given the shortage of labour to share such tasks with. One female key informant in Beresa *kebele* described her situation: *"my livelihood is not the same before and after I lost my husband three years ago. The burden of managing my home has doubled since the sole responsibility of undertaking productive and domestic tasks, as well as managing the education and health of our children, has rested on me alone."* FHHs are seriously affected by lack of time, since they have to stretch their time to undertake other productive and reproductive activities which are usually shared between husband, wife and children in MHHs.

Female heads see the time they spend on walking long distances to the market, and the long hours that collecting water demands, as a source of their poor economic performance. The trade-off of livelihood benefits in this case can be significant in terms of the pressure it creates on the allocation of time for other livelihood activities. For instance, when asked about the possible activities they would undertake if they were able to save the time they spend on fetching water and going to market, female FGD participants listed tending animals, compost building and even spending more time in the market. In view of these responses, it is not difficult to guess the impact of shortage of time on household production and domestic chores.

8.2.4 Access to credit

One of the mechanisms used to implement the Household Asset Building Programme (HABP) under the government's Food Security Programme is through the expansion of financial services in which Micro Finance Institutions (MFI) and Rural Saving and Credit Cooperatives (RUSACCO) are assumed to play a critical role (MoARD 2009c). In the study area, both formal community-based organizations (CBOs) such as MFI and RUSACCO, and informal organizations such as *ekub*[12] and *idir*, and individual money lenders, provide access to credit. Below, an attempt is made to present the gender disaggregated field results and discussions on access to these credit services and the prevalent gaps, and the implications of this with regard to household food security.

A. Access to formal micro credit services

Inquiries into whether respondent households are aware of the availability of formal micro credit services in their *kebeles* yielded a positive response by a large proportion of both female-headed and male-headed households. Yet, the figures, 67 percent against 82 percent for FHHs and MHHs respectively, show that female heads lag behind in receiving information. Various factors such as low educational level, low social capital such as an information network, and other such norms and perceptions can be barriers for women farmers in this regard. Out of the households who are already informed about the existence of formal credit service in their *kebele*, only 18 percent are beneficiaries. The findings generally show a low level of participation in both female- and male-headed households, the proportion of MHHs (nearly 19 percent) slightly exceeding that of FHHs (nearly 17 percent). See Figure 8.6.

---

[12] Community-based rotating saving and credit associations

Figure 8.6 Participation in formal credit services by type of household

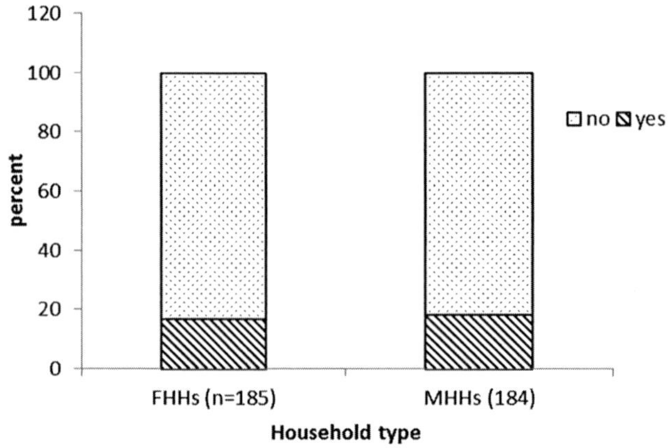

Source: Household survey, March to April 2011

FGDs held with both female- and male-headed households revealed that there is agreement about the benefits of credits in the livelihoods of rural households. Among these, the respondents underlined the importance of credit for purchasing inputs such as fertilizer and improved seeds. They underscored that credits act as a buffer against the dispossession of critical productive assets[13] for the sake of purchasing agricultural inputs. Moreover, households use credit for a wide range of purposes. These include payments of social obligations such as *idir*, health treatment costs, or the purchase of grain to fill food gaps. Thus, the role of credit is important as a coping mechanism at times of livelihood shocks and crises. However, the very low involvement of households in formal credit institutions regardless of gender, even though the benefit of credit is well known among the study population, requires further investigation.

Earlier study findings by Yigremew (2000) in Amhara and the Southern region showed that despite the claim made by regional Micro Finance Institutions (MFIs) in serving the rural poor and women, poor FHHs are mostly marginalized. This is because loans favour resourceful farmers rather than women with a low asset base. The survey finding above might not confirm the assertion by Yigremew at a glance. However, an expert interview held with an officer working for Omo micro finance institution[14], revealed that the collateral requirement is unfavourable to poor rural FHHs. Beneficiary eligibility depends on factors such as ownership of land and oxen, which FHHs mostly lack. Further discussions with female heads showed various factors behind their absence in Micro Finance Institutions (MFIs). Female heads in Beresa *kebele* associate their non-participation in formal MFIs with possible negative consequences such as loss of assets (oxen and land) in the case of default. Other factors, such as lack of financial management skills, lack of time to attend trainings, lack of interest and confidence, are also mentioned as possible reasons. The following short case stories of two female heads, one widowed and the other divorced, show some of the institution-related factors that influence the decision of FHHs to access credit services in the study area.

[13] Oxen and other livestock that are a very important form of buffer against food shortage.

[14] A government-owned regional organization operating in SNNP region. It has its seat in Meskan district. At the time of data collection it used to give members loans of up to 5000 birr (equivalent to 190€).

Case story 1

> *"I do petty trading in paper and traditional drinks for my livelihood. However, my business is severely suffering from lack of finance. Taking credit from private and government loan givers is not an option for me because I fear that I won't be able to pay back the loan. I fear that my children might land on the street if I can't pay back my debt."*

Case story 2

> *"I took credit, bought a donkey and started petty trade. I have worked for some time. I had much enthusiasm to work and to benefit from the petty trade. In the meantime, a new chairperson took over office and refused to sign me a paper so that I can take credit. He said I am too poor to repay the loan."*

In the above two stories, both women are well aware of the benefits of credit in terms of improving their livelihood. Yet, their decision to take a loan is severely constrained by their risk-aversive behaviour and by the existing institutional environment.[15] Here, two seemingly contradictory explanations might be seen as two distinctive constraints. In the case of the first story, the reason for the decision not to participate in credit services is the risk averseness of the loan taker, whereas in the case of the second story the external institutional environment acts as a barrier despite the willingness of the loan taker. The risk-averseness in this case emanates from an external institutional force which shapes the decisions and actions of individuals.

Unlike FHHs, MHHs did not mention collateral as a major issue in taking credit, which is a major departure from the position of FHHs. However, MHHs share the concerns of FHHs about the risks of default. In relation to this, the findings of the study revealed risk-averseness as a major reason for avoiding micro credit services. *"I know about Lakech Micro Finance Institution in my kebele. Despite the fact that I got a title as a 'model farmer' from the kebele, I am not a beneficiary of the micro credit service, mainly because I am not sure if I am capable of returning the loan, so I prefer to avoid it."*

The FGD discussion with male heads who are non-beneficiaries revealed that formal credit organizations mostly follow strict regulations and procedures when it comes to loan repayments. Thus, this has been a major constraint preventing many farmers from benefiting from the credit programme. A working system called "*yetelefa was*"[16] in a group-based loan-taking arrangement was identified by the respondents as the most unfavourable arrangement. In this kind of loan arrangement, not only do individuals find themselves at a higher risk of default and its negative consequences[17] but it also negatively impacts community-level relationships and cooperation. Information obtained from expert interviews in this regard revealed that defaulting is often associated with misconceptions on the part of the farmers and certain administration gaps. *"Some farmers considered loans as a grant from the government or simply as their right, which has facilitated the misuse of resources and the tendency to default. Rural communities used to give high value to the repayment of loans. In fact, the inappropriate disbursing of loans has contributed to the gradual erosion of this culture."*

---

[15] The response from the *kebele* official in the second case story shows institutional rigidness.

[16] An arrangement by which individual group members become guarantors for each other. It is considered risky for individual members if the others fail to repay their part for various reasons.

[17] One of the damaging effects is that defaulting households are not allowed to benefit from the fertilizer credit programme till they return their debt. During the second round field visit to the study area in 2012, however, FGD discussion with farmers revealed that such arrangements have been discontinued.

B. Access to Rural Saving and Credit Cooperatives (RUSACCOs)

The presence of rural saving and credit cooperatives is a very recent phenomenon in all of the study *kebeles*. Yet, at the time of data collection there were various economic cooperatives at different stages of their life cycle. Ega Rebi Saving and Credit Cooperative in Beresa *kebele*, a women's cooperative group organized around crop production in Dobo Tuto *kebele* (in its initial phase during the time of data collection), a women's cooperative group engaged in poultry raising in Semen Shershera *kebele*, a cooperative group organized by men around sand production in Dirama *kebele* and Dejen Saving and Credit Cooperative organized by a group of women around horticulture in Dirama *kebele* are some of the cooperative groups. The section below presents details of the Dejen Saving and Credit Cooperative and Ega Rebi Saving and Credit Cooperative in Beresa *kebele*.

Case study 1: Dejen saving and credit institution was established in the year 2008 in Dirama *kebele*. At the time of data collection, in the year 2011, the cooperative had 165 members. The cooperative is open to women above the age of 18, who have organized themselves to increase their income by engaging in horticulture production. Yet, depending on the individual's interest, the cooperative is also open to women who want to pursue various marketing ventures. The majority of its members are female heads. An interview with the head of the cooperative revealed some of the paths the cooperative has gone through in light of the challenges faced by its members, in order to enhance their economic capacity to improve their livelihoods.

In its initial phase the cooperative existed for two years with members saving 1 birr (0.039€) on a monthly basis. Gradually, this grew to 5 birr (0.19€) and finally to 12.50 birr (0.48€) by the time of the data collection. As the saving capacity of the cooperative grew, it got recognition from Netsanet Fana Saving and Credit Cooperative Union.[18] For the first time, it was granted a 86,190 birr loan by the Union in 2010 which was twice the amount of its capital at that time. The women who took loans were able to pursue petty trading, fattening of animals, marketing and other off-farm activities. However, the cooperative was unable to grow its saving capital and boost its potential in terms of loan size. As a result, the cooperative has gradually lost its members mainly because it could no longer satisfy the large demand for credit from its members. As a result, some of its members have gone to bigger credit providers and others have quit participating in credit schemes.

Case study 2: Ega Rebi, another saving and credit cooperative, is predominantly a men's cooperative in Beresa *kebele*. It initially started functioning with its members saving 2 birr on a monthly basis. Similar to Dejen, within its four-year initial period the cooperative lost many of its members. The management made certain reforms by allowing more flexible approaches to saving. Those who waited with patience have benefited in the long term as the cooperative increased its financial capacity. The cooperative eventually got recognition from Netsanet Fana Saving and Credit Cooperative Union. At the time of data collection, its members were able to borrow three times their savings. The minimum amount to start saving is 20 birr. A key informant who is one of the founding members explained "*We encourage those who want to join to start with a larger amount and we explain to them how they can realize their goal of seeing themselves financially strong within a short period of time*." The cooperative had 46 members in 2011.

---

[18] Netsanet Fana Saving and Credit Cooperative Union established in 2004 has its seat in Meskan district and provides loans for cooperative groups. The union is supervised by a zonal cooperative promotion office.

Comparing women's cooperatives in the study *kebeles* with other cooperatives that are predominantly run by men, it can be seen that there are various economic, institutional and administrative constraints that undermine the potential of women's cooperatives. One of the predicaments is related to government's lack of attention in building the capacity of women cooperative leaders which is reflected in certain administrative gaps these women's groups faced. In this regard, a female key informant said: "*Favouritism based on patronage relationships benefits men rather than women cooperative leaders. It is important to have good social connections and networks with local officials to receive on-the-job training.*" Moreover, it is evident from the above findings that women in the study area are less likely than their male counterparts to join lucrative income-generating ventures, because of lack of education and financial constraints. As a result, access to these cooperatives as well as the benefits accrued by them are skewed towards well-off male households. There is empirical evidence of this in the study area, such as a relatively well performing cooperative group made up of men who are engaged in sand production in Dirama *kebele*.[19]

Generally, even though these credit schemes have an immense potential to economically empower poor FHHs and contribute to their food security, serious gaps can be observed in the implementation process. As a result, these cooperatives are still unable to solve the financial problems of women to the extent demanded by their members. The withdrawal of members is a clear manifestation of such weakness. This in turn has both short term and long term impacts at individual and aggregate level. In the short term, by withdrawing from the service, women still remain income-constrained at household level, which has an immediate and direct effect on their household food security. Moreover, in terms of long-term consequences, it undermines the potential of these kinds of cooperatives to build a collective voice and solidarity among women. The latter could be a potential force to challenge biased gender power relations through enhanced group decision-making, an important prerequisite to break the institutional barriers which are a cause of women's vulnerability and food insecurity in the study area.

8.2.5 Access to agricultural extension services

The 2009 food security strategy document gives due attention to the transfer of knowledge based on a demand-driven agricultural extension service instead of the supply-oriented approach of the past. However, the reality on the ground shows that not all households from different socio-economic backgrounds are able to equally access and benefit from such services. Previously conducted empirical studies show that gender plays a critical role in such biases. For instance, Dejene (1994) asserts that extension services marginalize female farmers and as a result women are unable to access new technologies. Studies by Frank (1999) in the Amhara region also showed the low participation of women in agricultural extension training programmes, which led to situations in which their needs were largely ignored and side-lined. This is reinforced by the widespread perception in various regions of the country that being a farmer is synonymous with being a man. Below, the findings of the study regarding access to agricultural extension services are presented, with particular emphasis on contact with Development Agents (DAs), utilization of fertilizer and improved seeds.

A. Contact with development agents (DAs)

In order to assess whether there are gender differentials in terms of receiving expert advice, survey respondents were asked to rate the frequency of their contact with DAs in the year

[19] The cooperative has around 100 members. Even though eligible members should be young men under 30 years of age, such regulations are loosely practised in reality. The role of women in this cooperative is limited to providing catering service such as coffee and tea.

preceding the survey. Based on the survey finding presented in Table 8.2, a large proportion of both female-headed and male-headed households said that they had contact with the DAs in their *kebeles* "sometimes." Yet, it should be noted that the proportion of FHHs (nearly 66 percent) is greater than that of MHHs (58 percent). In the categories "often" and "very often", the survey finding showed a smaller proportion of FHHs as compared to MHHs: 32 percent and 9 percent of MHHs reported "often" and "very often" as compared to 21 percent and 5 percent of FHHs. FHHs are also disproportionately represented in the category of "no contact".

Table 8.2 Contact with Development Agents by type of household

Contact with DA	**Type of household**				**Total**	
	FHHs		**MHHs**			
	№	%	№	%	№	%
Not at all	15	8.2	3	1.6	18	4.9
Sometimes	120	65.6	106	57.6	226	61.6
Often	38	20.8	59	32.1	97	26.4
Very often	10	5.4	16	8.7	26	7.1
Total	183	100	184	100	367	100

Source: Household survey, March to April 2011

FGD discussions with both FHHs and MHHs showed that households generally felt that they were frequently receiving expert advice regarding various areas of their livelihood. Nevertheless, discussions with female-headed and male-headed households about the government-administered development programmes showed that male heads have a better understanding of these programmes than FHHs. For instance, FHHs were only able to name the Productive Safety Net and the family package programmes when asked to list the government programmes that are carried out in their *kebeles*. They have also very shallow and only very general information about the most important development interventions in their *kebeles*. In general, the discussions with both groups revealed that the tendency to seek information is higher among MHHs than FHHs. This might be due to the low level of literacy in FHHs. The other explanation for this gender gap in knowhow might be related to differentials in access to productive resources such as land and oxen. Due to their enhanced access to resources, male farmers enjoy the privilege of expert advice and follow-up, whereas women are overlooked as a result of their apparent absence in farming activities. These factors explain why MHHs are in a better position to access expert advice and information than FHHs.

With limited exposure to expert advice, the livelihood needs of FHHs remain neglected. To be more specific, the limited knowhow of women with regard to ongoing development interventions leads to the further marginalization of women from *kebele* meetings and trainings, which are mostly open to those active community members with a better knowledge of current affairs. In the long run, these circumstances facilitate the exclusion of women from development related decision-making process, although this would be vital for the improvement of their livelihood and food security.

B. Application of fertilizer and improved seeds

Chemical fertilizer

The survey finding on the amount of DAP fertilizer used in the production year 2010 is well-differentiated along gender lines. The finding presented in Figure 8.7 shows that nearly 54 percent of the respondents in FHHs, as against only 34 percent in MHHs, did not apply DAP. The proportion of respondents in both gender groups decreases as the amount of fertilizer increases, yet with a noticeable difference between FHHs and MHHs. The data summary for MHHs produced three potential outlier values which might inflate the average value for MHHs.[20] Thus, in order to avoid such biases, the outliers were deleted from the data. Accordingly, the aggregate average use of DAP in the 2010 maize production year in FHHs was found to be 11 kg as compared to 19.4 kg in MHHs. When the average use of DAP per unit of land is computed for each household category, the figure for FHHs is 5.51 kg/*timad* (22.04 kg/ha) and 7.95 kg/*timad* (31.83 kg/ha) for MHHs. As compared to the national figures of CSA (2011, 9-11) the application rate of chemical fertilizer for maize and sorghum for the year 2010/11 which is also the highest and the lowest among five types of cereal crops[21], was 175 kg/ha and 97 kg/ha respectively. Compared to these figures, the findings generally show that smallholder agriculture in the study *kebeles* is characterized by extremely low chemical fertilizer application. Moreover, gender wise, lands managed by FHHs receive less chemical fertilizer than those of MHHs.

Figure 8.7 Amount of DAP fertilizer used by female- and male-headed households in the production year 2010

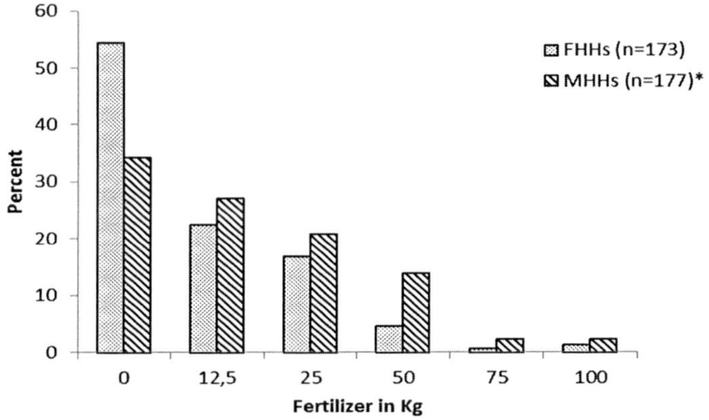

* *The total excludes three outliers, each with a value equal to 1000Kg*

Source: Household survey, March to April 2011

Qualitative information obtained from the focus group discussions clearly shows that the use of inputs such as chemical fertilizer or improved maize seeds is widely practised in all of the study *kebeles*. Participants in FGDs in both female- and male-headed households unanimously agreed on the benefits of inputs such as fertilizer and improved seeds in terms of increasing the yield on their land. In fact, without such inputs, poor harvests are considered inevitable as farmers speak of their land becoming increasingly dependent on DAP and urea.

[20] There is a standard deviation of 81.4 in MHHs as compared to 16.9 in female-headed households when the three outlier values are included.

[21] *Teff*, barley, wheat, maize and sorghum

Improved seeds

In addition to chemical fertilizer, improved seed is another input that contributes to an increase of agricultural productivity. Even though the amount of improved seed and the extent of area under application are increasing from year to year at national level, use of these improved seeds still remains very low and has not been widely practised by smallholders (CSA 2011). The gender disaggregated result of the survey in the study *kebeles* shows 73 percent of FHHs and 53 percent of MHHs did not use improved seed during the 2010 production year, thus showing a greater proportion of FHHs.

Opinions of farm households regarding trends in agricultural input (chemical fertilizer or improved seed) use and the gendered implications

Survey respondents were asked if they noticed any increase, decrease or no change in the use of agricultural inputs (either chemical fertilizer or improved seed) over a period of five years before the year of data collection. Computing the proportion of households who noticed a production decrease (excluding those who reported no decrease and those who did not respond), nearly 42 percent of FHHs and 48 percent of MHHs noticed a decrease in application of inputs.[22] Even though it is hard to tell whether the findings in the study area showed a negative or positive scenario (since 58 percent and 52 percent of FHHs and MHHS did not respond) when compared with the findings at national level, it is worth focusing on the reasons given by households who noticed a decrease in input use, in order to get some insights into the likely response of the other respondents. However, the question regarding the reasons for input decrease, has similarly failed to get a response from a large proportion of both female-headed and male-headed households (Table 8.3).[23]

Table 8.3 Reasons for the decrease in agricultural input use by type of household

Reason for decrease in input use	**Type of household**				**Total**	
	FHHs		**MHHs**			
	№	%	№	%	№	%
High price	51	27.4	66	35.5	117	31.5
Lack of information	17	9.1	13	7.0	30	8.1
Lack of interest	5	2.7	5	2.7	10	2.7
Others	5	2.7	6	3.2	11	2.9
No decrease	33	17.8	49	26.3	82	22.0
No response	75	40.3	47	25.3	122	32.8
Total	186	100	186	100	372	100

Source: Household survey, March to April 2011

The single most determinant factor for low input use is the associated high cost of agricultural inputs accounting for 27 percent of FHHs and 36 percent of MHHs. All qualitative information obtained through separate FGDs with MHHs and FHHs supports the above findings, emphasizing that the use of inputs such as fertilizer and improved seed is becoming less af-

[22] The proportion of female and male respondents who reported production decrease is calculated by adding the number of respondents who identified reasons for decrease in input use in both categories.

[23] Informal discussions with the enumerators revealed that the reasons behind this were different from household to household. Among them were failure to recall memories and unwillingness for privacy reasons.

fordable through time. This is followed by poor accessibility of information with nearly 9 percent and 7 percent of female- and male-headed households respectively.

To further strengthen the above quantitative findings, qualitative inquiry around the issue of fertilizer use among farmers in the study area revealed certain social attributes in respect of gender differentials in terms of access to agricultural inputs, decision-making power and impact on food security. The section below draws on information obtained from interviews with a man and a woman farmer in the study area on their reaction to a newly introduced system of fertilizer allocation that was put in place by the *kebele* input distribution task force in the year 2012.[24]

> "*The kebele officials put a lot of pressure on the farmer to make a one-time full payment for DAP and urea. Nevertheless, from my experience, the land requires urea at a later stage of the crop's life cycle than DAP. Therefore, I asked the kebele to allow me to pay separately at different times instead of selling my goats to pay for DAP and urea at once. Nevertheless, they refused to accept my offer. I know very well that it is my right to go and complain at district level, so I took the case there. I got a positive response from the district, they were even willing to offer me DAP on spot. When the kebele people got the information that the district is about to accept my offer, they pleaded with me to buy DAP at the kebele office since they did not want to forfeit their benefit from selling DAP. Many in my community, who do not have adequate knowhow about their rights, simply turn to some other risky options that eventually damage their livelihood. For instance, they limit the amount of fertilizer they buy to one to two timad and as a result their production becomes meagre. As a result they are not able to benefit from the input programme that the government offers. I shared my experience with my neighbours and some of them were able to benefit.*"
>
> *A male farmer in* Beresa *kebele*

The above empirical evidence highlights the nature of social interactions between external local actors and farmers during social encounters and processes of acquiring access to chemical fertilizer. The experience of the farmer in the case story shows the extent to which farmers are able to influence the institutional environment through their decisions and actions. It shows how structural processes are shaped through the agency of farmers in the study area. From the farmer's response, it is clear that the agency of farmers at a given time and location depends on their knowledge and articulation of the external institutional environment with which they are confronted in their day-to-day life. Yet, their ability to make a difference depends on how well they are able to interpret the information they receive from their social environment and their capacity to engage themselves in a practical or discursive type of discourse.

In order to cope with the hassles of the bureaucratic structures and achieve his objective, the farmer has to go through certain social interactions and processes. To achieve his objective, which is to gain access to fertilizer, he has to deploy his knowledge of the rules and regulations and processes that dictate interaction both at *kebele* and district level. Moreover, positive behavioural attributes, such as his ability to understand the existing social, economic and political setting, his bargaining and negotiation abilities, as well as his confidence and self-assertiveness, are important pre-requisites that helped him to confront the existing external

[24] Focus group discussions with farmers during the second round data collection in May 2012 revealed that a new system of fertilizer administration had been established to replace the system of fertilizer provision through deposit and credit (25 percent to 50 percent deposit). The new system requires farmers to make full cash payment upon purchase.

structural rules and processes. Generally, his proactive engagement within the existing institutional setting led to certain results in favour of his livelihood needs.

On a similar issue of fertilizer acquisition, a widow in DoboTuto responded with this story:

> *"I know well that fertilizers and improved seeds are important to the productivity of my land. But I am too poor to purchase fertilizer even with payment in two instalments. Taking credit is not an option for me since I fear that my production will not be enough to pay back my debt."*

When asked if she went to the *kebele* and district levels to ask for a solution, like the man in the story above did, she responded;

> *"Because I am a female head, I feel I can't do that, I will not go that far. It would be like clapping with one hand, no one will hear me. Therefore, the only option left for me is to feel pity for myself. Male-headed households have a better chance than a widow like me, since they have various means to get credit and they are the credit-worthy ones. But as a lone mother no one is willing to give me credit."*

The response of the widow shows how unequal gender power relations play out in the social field to negatively affect people's livelihoods with regard to benefits such as access to fertilizer and thus enhanced productivity. One of the ways in which such inequality manifests itself, according to the response above, is through differential access to economic assets. For instance, even though the use of inputs is equally desirable for both FHHs and MHHs, its access was restricted for the respondent above due to her poor purchasing power or her low asset base. In the face of extreme need, women farmers may resort to the disposal of other assets that serve as a buffer in times of crises, thus putting their food security at greater risk. Unequal gender power relations also become evident through the weak bargaining power on the part of the respondent widow. It is primarily the influence of various norm- and tradition-based values and perceptions that dictate the submissiveness of women in such circumstances.

The local institutional environment forces women to live with the existing situation instead of encouraging them to use their agency to effect changes. In addition to this, lack of education and lack of access to information and other resources compound together to negatively impact their decision-making ability. As a result of all these factors, the decisions made by poor FHHs under such circumstance remains at a level of practical consciousness which leads to the confirmation or reproduction of the existing institutional structures. Generally it can be said that structural changes that are introduced through external interventions can benefit one group while disadvantaging the other. Yet, simply accepting or confirming such changes might pose formidable challenges in the long run. For already vulnerable groups such as FHHs, this might translate into greater vulnerability to livelihood shocks, and a weak capacity to cope, thereby threatening their household's food security.

8.2.6 Access to water infrastructures

Public infrastructures such as roads, market places, irrigation, water schemes, veterinary services, education and health play a very important role in food security by enhancing the productivity of labour and land, by facilitating commercialization and reducing transaction costs (Tesfaye and Degefa 2009). Among these, this section selectively deals with access to safe drinking water as one critical determining component of the food security of farm households in the study area. In addition to its domestic use for drinking, cleaning and cooking, households are heavily dependent on water for activities such as livestock rearing and crop production. Table 8.4 provides an overview of the distribution of water sources and services in each of the study *kebeles*.

Table 8.4 Distribution of water infrastructure in the study *kebeles*

Name of the study *kebele*	**Source of potable water**		**Irrigation**	**Water harvesting ponds**
	Functioning water points	Natural sources		
Beresa	3 & few private	1 river	Yes	Yes
Dobo Tuto	2	-	-	<5
Dirama	1	1 river outside the *kebele*	-	Yes
Semen Shershera	2	1 river outside the *kebele*	-	15

Source: own compilation during field work, 2011

Beresa *kebele* is located adjacent to Dobo Tuto *kebele* and has a total of 3 water points serving the entire *kebele* by the time of data collection. The location of the water points does not provide equal accessibility for the entire villagers in the *kebele*. The large influx of people coming from neighbouring *kebeles* to these water points further constrains access for the community of Beresa *kebele*. Villagers speak of frequent encounters with long queues of around 400 people with dozens of donkeys waiting at a single water point. During the time of the data collection, all of the water points in Beresa *kebele* were having technical problems, which require households to travel to the nearby town. The Irinzaf River in Beresa is a common spot for cattle. Lands that are close to the river benefit from irrigation. There are also a few water harvesting holes that are partly functional.

Dobo Tuto *kebele* is a water-stressed *kebele* by its nature. Potential rivers in the *kebele* have dried out due to various environmental and human-related factors. There are two water points serving the entire *kebele*. Areas which are below the mountain have better access to clean water than those who are above. Pointing out the technical problems associated with the few available water points, villagers expressed high concern for their livelihoods as a result of the chronic scarcity of water.[25] Households pay 6 birr (equivalent to 25 cents) for two pots of water on a daily basis. In extreme situations, villagers are forced to seek water points in the neighbouring Beresa *kebele*, yet the poor road network is another constraint on easy movement between the *kebeles*. Although water harvesting is considered a helpful technological practice for the provision of water for domestic water needs, there are only a few ponds in the *kebele* and these are mostly accessible only to model farmers. Figure 8.8 shows pictures of women performing their daily routine of fetching water in the study *kebeles*.

[25] Poor engineering in the construction of water points is considered a major cause of the poor service.

Figure 8.8 Water sources and women traveling to haul water

Source: Field work, 2011

Dirama *kebele* has one water point. The water point is located in a hilly area which makes transporting water an extremely difficult task. As a result, households often travel to the nearby town of Butajira to fetch water. Depending on the location of their village, the round trip to Butajira town to fetch water may take up to 5 hours. There is a serious shortage of potable water both for domestic consumption and for crop cultivation and livestock keeping. At times of severe shortage of water, households use a river called Akamuja for drinking and other purposes, located outside the *kebele*. This is despite the serious health risks associated with it. A plastic container of up to 20 litres costs 20 cents in the nearby town of Butajira.

In Semen Shershera, there are two boreholes dug by an NGO that are now administered by the government in the *kebele*. The two water points serve the entire households in the *kebele*. Households also use the Akamuja River by crossing the neighbouring *kebele*. In addition to these, there are around 15 water harvesting holes that are considered functional according to an interview held with a natural resource management expert. However, the response from the farmers in this regard casts some doubt on the practical applicability of the water harvesting holes.

So far, an attempt has been made to show the distribution of the available sources of water in the *kebeles,* revealing a scarcity of water in all of the study *kebeles*. In all of the four study *kebeles,* the availability of potable water for domestic purpose is a problem and rain harvesting and irrigation technologies are poorly developed. The following section focuses on the findings of the study regarding the gendered impacts of potable water scarcity at household level. Secondly, it presents the findings in respect of access to rain water harvesting technology in the study *kebeles.*

Gendered impacts of lack of access to safe water on food security

A gender-based assessment of the impact of scarcity of water on food security in the study area showed that FHHs face severe challenges primarily due to time and labour constraints. For instance, due to their gender roles, women spend a substantial amount of their time hauling water, which constrains their participation in other productive activities (see also 8.2.3). Easier access to safe water reduces the time spent hauling water by women, thereby increasing their productivity and their economic status. Yet, in the absence of this, productivity in FHHs is compromised, resulting in increased pressure on food provision at household level. Likewise, the strong linkage between the availability of water and food security is well demonstrated in previous studies made in different regions of the country (Tesfaye 2012; Tesfaye and Degefa 2009).

Further qualitative investigation through expert interviews revealed that poor access to potable water has a substantial health impact both on humans and livestock. This is primarily because the lack of access to safe drinking water forces poor households to resort to water sources with potential health risks. For instance, in the face of extreme water scarcity, households in Beresa *kebele* depend on the Irinzaf River for water for human and livestock consumption. Similar situations are also reported in Semen Shershera *kebele* where farm households resort to the Erensha (Akamuja) River for drinking and household consumption. In these cases, both livestock and humans are exposed to water-borne infectious diseases. An interview with a health extension expert in Dirama *kebele* reveals that goitre and malaria are diseases that commonly affect women in the study *kebeles*. The extra cost of medication in turn creates additional financial stress for poor FHHs who already have limited income sources. Illness affects the intake and utilization of food and affects their ability to engage in productive activities, with serious repercussions on their food security and well-being.

Gender gap in access to water harvesting technologies: the case of rain harvesting ponds

To overcome the pressure created by erratic rainfall and scarcity of water in rural areas of the country like the study area, and with the aim of improving food security in areas of highly variable rainfall, the development of water resources in rural regions of the country is given due attention as part of the rural development strategy of the government. Accordingly, several rain water harvesting ponds have been constructed by regional states, NGOs, communities and individual farmers throughout the country.

The study *kebeles* are among those water-stressed *kebeles* that have been selected to benefit from the introduction of the water harvesting technology. Even though the intervention is relatively new in the study area, expert interviews in Semen Shershera *kebele*, for instance, revealed that there are further plans to expand the technology under the safety net program in the *kebele*. Against this backdrop, this section presents the study findings regarding access to such technology by female- and male-headed households, in order to show if there are gender gaps in benefit sharing.

Explaining the conditions of access to rain harvesting technology, one female participant expressed her disillusionment as follows: "*Even if technology comes, it comes only for households where there is a man, our gender has limited us.*" Her response speaks loudly about the existing inequality in terms of access to rain harvesting technology in the study area. Further investigation in Beresa *kebele* revealed that water harvesting appliances were freely distributed by extension workers to those individuals who are capable of excavating in their yard and to those willing to benefit from it. The situation in the other three *kebeles* shows a similar pattern. Such activities are associated with time, cost and labour and thus are most likely to be managed by labour-abundant households with good financial resources. Thus, such an offer is more likely to benefit wealthy and labour-abundant MHHs, while it systematically marginalizes weak and elderly female-headed households with poor ability to mobilize labour. In the context of this study, personal observation in all of the study *kebeles* confirmed that the existing water harvesting facilities are in the yards of MHHs who have been honoured with the title of "model farmers". The findings of the study agree with Rebeka's (2006) finding in Alaba district in which lack of time, knowhow, and preference for male farmers by agricultural extension experts are identified as reasons that restrict women to benefit from water harvesting technologies.

Generally, technologies such as water collecting appliances, as important as they are to the livelihood of FHHs, should be supported by a detailed study and need assessment so as to ensure that the benefits are accessible to all. This means including the less able and the resource-poor FHHs. If such technological interventions are not tailored to meet the needs of women farmers who play an essential role in food provision, solving the water shortage problem will only be a partial success. Thus, in order to balance the burden imposed on women due to gender roles in the study area, intervention in the form of labour- and time-saving technologies need to be available and accessible to women.

Even though the purpose here is not to discuss the pros and cons of the water harvesting intervention in the study area, the field visits to the study *kebeles* revealed certain gaps in its implementation which can be roughly described as technical- and knowhow-related or administrative constraints. Lack of expert follow-up, poor quality of plastic distributed to farmers, lack of proper fences, and accumulation of water for long periods are some of the observed problems. As a result, some households are not convinced by the functionality of the technology *per se* and suggest other innovative[26] ways to harvest rain water which they consider more efficient in terms of sustainable use. Further investigation is required into the question whether there are other technologies that would fit the needs of both gender groups, or whether the existing technology can be improved to meet the pressing needs of FHHs.

8.3 Decision-making and food insecurity

The livelihood strategies that farm households pursue towards ensuring their food security are based on the various decisions they make in their day-to-day routines. In vulnerable livelihood systems such as those studied here, the patterns of decision-making are also subject to certain transformations. Thus, gender-based micro-level studies regarding inter-household decision-making on the subject should be an important part of livelihood studies, since they provide key information for policymakers so as to tailor development programmes to the needs of both female- and male-headed households. Researches done to capture the dynamicity of decision-making are very limited in Ethiopia. The ones available primarily focus on the existing patterns of intra-household decision-making. This section aims to capture the existing trend in decision-making processes among the study population by taking female- and male-headed households as a major unit of analysis. In order to achieve this, the author believes that no conventional standard questionnaire is able to go as deep as uncovering the real practices, decisions, and the existing trends through time at household level and beyond. Accordingly, the study depends on views and perceptions of farmers captured through focus group discussions and key informant interviews. The aim is to show possible links between the nature of decision-making and the food security situation of female- and male-headed households.

[26] For instance building water storage tanks above the ground connected with pipes coming from the roofs of houses during the rainy seasons.

Decision making in male-headed households

In general, major livelihood decisions in MHHs include what to produce, what to sell and what to use for household consumption, how and when to sell, how to produce, what to purchase and how to distribute resources. Decisions such as what to produce and which produce to use for household consumption and which for sale are usually determined jointly. For instance, the decision to use maize and sorghum for consumption and *teff* and haricot beans for sale is reached through discussion between the husband and wife. Men are usually in charge of selling high-value products such as *teff*. Following the local gender-based norm, the selling of livestock such as oxen, sheep and goats is also a men's domain. Women are active in the exchange of low-value goods such as vegetables, and animal products such as butter and eggs. In addition to these, depending on household demand, decisions on how to distribute resources among various need areas involves the husband and wife and grown-up children, if there are any living in the house. For instance, family members participate in decisions regarding allocation of money for the purchase of fertilizer, purchase of improved seeds or payment for school fees or rent.

The question put to MHHs regarding existing trends in decision-making on household affairs got quite diverse responses depending on the socio-economic characteristics of the household, individual behaviour, level of education and age of the household head. Despite such heterogeneity, the findings generally show two seemingly contradictory trends. On the one hand, the participants made a notable remark by emphasizing the gradual transformation in the ways decisions are made within the household. Earlier, the entire decision-making process, including the final decision, used to be controlled by the male household head. Husbands had the final say concerning various important matters, including the fate of his children and that of his wife. Yet, in the last 5 to 6 years, decisions have begun to be made on a participatory basis, involving every member of the household including wives and children.

Households mentioned various reasons behind such shifts. Some attribute it to the awareness-creation programmes regarding the roles and rights of women, offered by NGOs such as Progynist which have been active in some of the study *kebeles* in the last couple of years. Others argued that external environmental pressures[27] and the subsequent economic downturn that hit most farm households has facilitated the process.
To further elaborate this, the statement by one of the FGD participants presented below shows how past economic crises have contributed to shaping thinking and decisions and practices in MHHs.

> *"Poverty and our coping experience in the past has taught us the advantage of thinking and acting together based on the principle of sharing, rather than doing things in a self-centred manner under the pretext of the so-called 'men's business'. All of us know how we escaped from bad times through the unlimited role of our women. What we learned from the past is nothing but the positive impact one can make by shifting responsibility and sharing obligations to alleviate hardships."*
>
> *A male head from Beresa Kebele*

On the other hand, generalized statements like the one given above might fail to capture some details on the ground. Further inquiries revealed that, except in certain areas, the role of wives in the final decision is still limited to areas which by tradition are considered to be a women's

[27] For instance external factors such as shortage of water, and the subsequent pressure on time, have disturbed daily routines and therefore altered the pattern of decision-making.

domain. A male key informant household head said: *"Quite different from the earlier years, what makes the nature of the decision-making process inclusive and more participatory is the fact that the man these days lets his wife and children know his plans. However, final decisions are still made by the man in important livelihood matters."* According to this opinion, it is not the case that children or wives participate in making final decisions in all livelihood areas. Some decisions, particularly those with a high livelihood value, are left to the man who is considered to be the head of the household. Yet, there are still some exceptional cases in which grown-up children actively participate in negotiations, problem-solving and final decision-making.

Decision-making in female-headed households

FHHs exercise full self-autonomy in many areas of their livelihood. The following two comments made by a divorced and a widowed female head during a FGD session describe the nature of this independent decision-making female heads enjoy in managing their livelihoods.

> *"As a divorced woman, I enjoy more freedom now than when I was married. Now I am capable of paying every expense in my household, including idir fees, tax, fertilizer and improved seeds as well as other domestic expenses. I feel I am more independent."*
>
> *A divorced woman*

> *"I consider myself self-sufficient since I am able to do most of the tasks by myself except things such as ploughing and selling cattle, in which my son or a relative is sometimes involved. When I was married, I was not able to decide on important issues such as selling goats or grain. My deceased husband used to decide what and when should be sold completely on his own."*
>
> *A widow*

Female FGD participants agreed on the following characteristics of a self-sufficient/ independent/ empowered female head:

> *"The one who is able to secure food on her own and buy clothes for her children, she is able to pay for social expenses such as idir fees as well as government debts in her household. In addition, she buys oxen and cows, hires a farmer, and sends her children to school, and constructs houses if she doesn't have debts."*

The first two responses show that female heads enjoy greater self-autonomy in decision-making as compared to women in male-headed households. These decisions are made around livelihood undertakings such as domestic chores, farm duties, participation in non-agricultural activities, as well as income distribution/ cash spending among various livelihood needs. What the responses above do not show is the level and nature of the decisions female heads are able to make in social encounters at the extra-household level. Yet, in terms of their relevance to food security, such aspects deserve attention.

Despite the above responses, there are some areas where women farmers' ability to make decisions faces stiff challenges as a result of the existing gender-based structures, norms and societal perceptions. As discussed in the previous chapters, it is very important for female heads to conform to the characteristic attributes of a "good woman", regardless of the low social status they have in their community. For instance, in FHHs gender plays an important role in shaping the kind of decisions made by women regardless of the absence of a male figure. Important activities such as ploughing in FHHs require the involvement of male relatives or male community members. Given their very low social and economic status, they usually make decisions from a lower power position regarding what to grow, when to plough, what and how much inputs to use and how to do the follow-up tasks. Their ability to negotiate and make favourable livelihood decisions determines the size of their harvest, with considerable effects on the availability of food at household level.

Further insights into the issue under discussion can be drawn from the response of a female head FGD participant in Dirama *kebele*:

"A *female head in our village is perceived as one who does not have a controller, that is a male figure in her house. Calling of names such as "tekochi yelelat" meaning one who has no controller, is something that we are very familiar with. Society assumes a woman who does not have a husband belongs to everybody. Therefore, a good woman is one who keeps her dignity; who does not bring shame to her community by spending nights everywhere, or who never stays out late at night."*

This response shows that in conforming to the existing structures, female heads compromise their opportunity to exercise decisions in certain areas of their livelihood. Generally, from the findings so far, it can be argued that the power of FHHs to make decisions at the household and extra-household levels cannot be understood separately from the social and economic position they are given in their community. And in turn their food security cannot be detached from the level and nature of decisions they are able to make.

8.4 Interim summary

The study hypothesized that with less access to resources and decision making powers, FHHs are more food insecure than their male counter parts. Accordingly, this chapter revealed that food insecurity is a common problem of both FHHs and MHHs in the study *kebeles*. Yet, each year FHHs face greater food gaps than their male counterparts. A greater proportion of FHHs (nearly 77 percent) identified themselves as seasonally food insecure than male-headed households (nearly 65 percent). Various factors that have their roots in the existing gender-based norms, values and customary practices prevent FHHs from accessing important livelihood assets, which translates into their relatively restricted decision-making power at extra-household level. In this regard, the study shows that FHHs have smaller plots of land (0.5 ha) and enjoy less control over it as compared to their male counterparts (0.7 ha). Nearly 52 percent of FHHs against 40 percent of MHHs do not possess oxen. Shortage of labour and time to engage in various on-farm and off-farm agricultural activities are another constraint. The existing policy environment is more favourable than in the past in terms of access to credit for FHHs, and their involvement in saving and credit cooperatives is significant in all of the study *kebeles*. Yet, FHHs are involved in less lucrative and small-scale income-generating activities as compared to their male counterparts, so that the benefit they gain from the credit services is limited. In terms of access to agricultural extension services, the study shows that male heads have better contact with agricultural experts than female heads. Moreover, with regard to the use of agricultural technologies such as fertilizer the study revealed that nearly 54 percent of the respondents in FHHs, as against only 34 percent in MHHs did not apply DAP for the production year 2010. Of those who apply fertilizer, a larger average value per *timad* is applied on lands operated by MHHs. Lack of access to potable water and water infrastructures imposes unique challenges on FHHs, primarily due to the existing gender roles and constraints of labour and time. In relation to decision-making, FHHs exercise full autonomy in areas where their traditional gender roles give them adequate legitimacy. Yet, in livelihood spheres that by tradition are assumed to be "male domains", or during social encounters where negotiations and bargaining with their male counterparts become inevitable, their decision-making is compromised due to the existing gender-based structures. Generally, as a result of their lack of access to productive resources and their weaker position in decision-making, FHHs are more vulnerable to food insecurity as compared to their male counterparts. Thus, the findings generally confirm to the presumption made.

9 Coping strategies against livelihood shocks and stresses and the weakening traditional coping mechanisms

Various traditional coping mechanisms have for long been practised by the study population with the aim of achieving food security at household level. Yet, the increasing frequency of weather events and environmental changes, the gradual social transformations over the years[1] and the subsequent effects on livelihood resources, continue to undermine the existing traditional coping mechanisms. One of the manifestations of weak coping mechanisms is the persistence of seasonal food insecurity that characterizes the livelihood of the majority of farm households in general and that of FHHs in particular.[2] As a result of this, new coping strategies are becoming increasingly important to the daily survival of the study population who are met by the unprecedented livelihood shocks and stresses.[3]

The study hypothesized that female-headed households have weaker coping strategies against livelihood shocks and stresses as compared to their male counter parts. The results of the investigation are presented in three main sub-sections. The first section deals with households' experience of livelihood shocks and stresses. This is followed by a summary of the type of traditional coping mechanisms and coping strategies employed by households in the study *kebeles*. The third sub-section presents the findings of the study on response mechanisms employed by households against income stresses. Throughout the chapter an attempt is made to show how coping with livelihood stresses and shocks plays out in the social field, by putting the role of guiding institutions and resources[4] available to households at the centre of the discussion. This chapter shows how unequal gender power relation shapes the actions of men and women farmers in their struggle to sustain food security and other well-being needs amidst unusual shocks and stresses.

9.1 Livelihood shocks and stresses

Within the definition of the SLF, the ability to cope with and recover from stresses, as well as the ability to maintain and enhance capabilities, are the basis for a sustainable livelihood (Scoones 1998; Chambers and Conway 1992). This definition puts "shocks" and "stresses" and the nature of coping strategies at the centre of livelihood analysis. Accordingly, this section begins with an investigation into the types of shocks and stresses that are experienced by farm households, using the information gathered through household surveys, FGDs and key informant interviews.

9.1.1 Frequency of Shocks

The survey respondents were asked to recall the number of times they experienced severe shocks within the five years preceding the year of data collection. It was found that around 53 percent of FHHs experienced two shocks, as compared to 34 percent of MHHs. A large proportion of MHHs (35 percent), almost double the figure for FHHs (13 percent), reported they experienced livelihood shocks more than three times during the five-year period (Figure 9.1).

[1] Farmers' perceptions regarding environmental and social changes are presented in section 7.1.

[2] This aspect is discussed in chapter 8. See Figure 8.3

[3] For the definition of coping mechanisms and coping strategies see sub-section 2.1 pp.11

[4] The gender dimension of access to resources is discussed in chapter 8.

Figure 9.1 The frequency of shocks experienced by type of household from the year 2005-2010

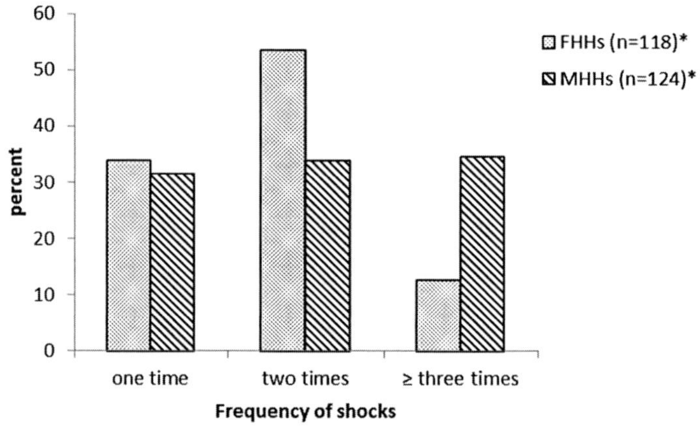

*All of the respondents have reported that they have experienced some kind of shock in the given period. Yet, a large number of both FHHs and MHHs did not specify the number of times they experienced the shocks in the given period

Source: Household survey, March to April 2011

Drawing conclusions from the above figures about the possible impacts of shock events can be erroneous. Following the argument in Dercon (2000) as well as Gertler and Gruber (2002) regarding the nature, frequency of shocks and the ability of people to recover from them, the fact that a large proportion of FHHs in the study area have experienced shocks only once or twice, or in the case of MHHs more than three times, does not tell us much about the severity and the possible consequences of the shock. The only conclusion that can be drawn here is that shocks are not unique to a specific type of household, but are experienced by both FHHs and MHHs in the study area. Thus, it can be argued that further investigation into the nature of these shocks is important in order to make a comparative analysis of the impact of these incidents on the coping capacity of female- and male-headed households.

9.1.2 Nature of livelihood shocks

In the study area, both common and idiosyncratic risks act together, putting serious pressure on the food security of households as well as their ability to cope. Thus, distinguishing between the two is often a challenging task. The findings below explain this further.

The survey attempted to gather data on types of shocks encountered by households within the year proceeding the year of the data collection (2010). As illustrated in Table 9.1, a larger proportion of FHHs as compared to MHHs faced two or more than two types of shocks in the year 2010. For example a large proportion of FHHs witnessed illness of a family member. Combined with other forms of shocks, the proportion of those who reported illness of a family member is found to be 32 percent. In the case of MHHs, illness of a family member combined with other shocks made up only 25 percent, but still took the first place among other forms of shock. The survey also showed that 27 percent of and 21 percent of FHHs respectively reported death of livestock and crop failure as the second and third most frequently encountered shocks. In the case of MHHs, the second and third places are occupied by death of livestock and shortage of income, each accounting for 23 percent and 21 percent respectively.

Table 9.1 Shocks experienced in the year 2010 by type of household

Type of shock	Type of household				Total	
	FHHs		MHHs			
	№	%	№	%	№	%
Illness of a family member	20	12.1	25	15.3	45	13.7
Death of a family member	12	7.3	4	2.5	16	4.9
Death of livestock	24	14.5	24	14.7	48	14.6
Crop failure	27	16.4	31	19.0	58	17.7
Shortage of income	11	6.7	37	22.7	48	14.6
Illness of a family member + death of livestock	17	10.3	10	6.1	27	8.2
Illness of a family member + death of livestock + crop failure	1	0.6	3	1.9	4	1.2
Illness of a family member + shortage of income	15	9.1	2	1.2	17	5.2
Death of a family member + death of livestock	3	1.8	1	0.6	4	1.2
No shock	35	21.2	26	15.9	61	19.0
Total	165	100	163	100	328	100

∑Illness of a family member in FHHs=53 (32.1 percent), MHHs=40 (24.5 percent)
∑Death of livestock in FHHs = 45 (27.3 percent), MHHs= 38 (23.3)
∑ Crop failure in FHHs=28 (17 percent), MHHs=34 (20.9 percent)

Source: Household survey, March to April 2011

Information gathered through FGDs has some consistency with the survey result above. Accordingly, illness of a family member, death of domestic animals, crop failure and shortage of income are the most serious and highly prevalent forms of shock in both female- and male-headed households. The following sections present details on crop failure and death of domestic animals. [5] In the case of death of domestic animals, the section specifically discusses about livestock disease since further quantitative inquiry and qualitative information clearly revealed that livestock disease is the primary cause of death of livestock in the study area.

Crop failure

A majority of the surveyed respondents (63 percent in FHHs and 72 percent in MHHs) reported that they had observed a decrease in production in the year 2010 as compared to 2009. During the focus group discussion, farm households very often spoke about not producing enough maize, the main staple crop in the study area. One model male participant pointed out that he had witnessed a constant reduction of productivity on his farm land over the years. *"For last year, I expected that I would get 6 quintals of maize per timad of my land.*[6] *However, I got only one quintal. Again, in the same year, I was expecting the productivity for teff to be 4 quintals per timad, but what I got was only 1 quintal per timad."* For some households,

[5] Illness among humans is regarded here as an idiosyncratic factor and is thus not discussed here as it is beyond the scope of the thesis.
[6] He explained that under good rainfall conditions he used to produce up to 20 quintals of maize per *timad*. This implies that under the present conditions his productivity is only 1/3 of the productivity potential of his land.

the problem was even as worse as a complete harvest failure. Summarizing the responses obtained, three major factors stand out as causing low crop yield in all of the study *kebeles*. These are categorized as disease infestation, rainfall irregularity and problems related to agricultural inputs.

Equal proportion of female- and male-headed households that account to 20 percent of the total respondents in each category reported no decrease of maize harvest in the production year preceding year of the survey. Among those who reported a production decrease, large proportion of both female- and male-headed households reported rainfall irregularity as the major cause for the poor maize harvest. This accounts for nearly 43 percent of FHHs and 38 percent of MHHs. Around 37 percent of MHHs identified crop disease as the cause of low production against only 12 percent of FHHs. Agricultural input related constraints are the second largest reported reason accounting to 38 percent (Figure 9.2). In connection with input-related constraints, a larger proportion of FHHs relates these to their limited use of improved seeds and fertilizer, primarily due to financial constraints. In the case of MHHs, the information obtained through FG discussions and in-depth interviews with DAs show that crop diseases are primarily caused by frequent defects in the improved seed varieties of maize. Yet, there is a strong assumption on the part of the farmers that local weather changes, such as the irregularity of the *belg* rains, have increased the susceptibility of crops to diseases and pests.

Figure 9.2 Cause of low maize yield by type of household that reported production decrease

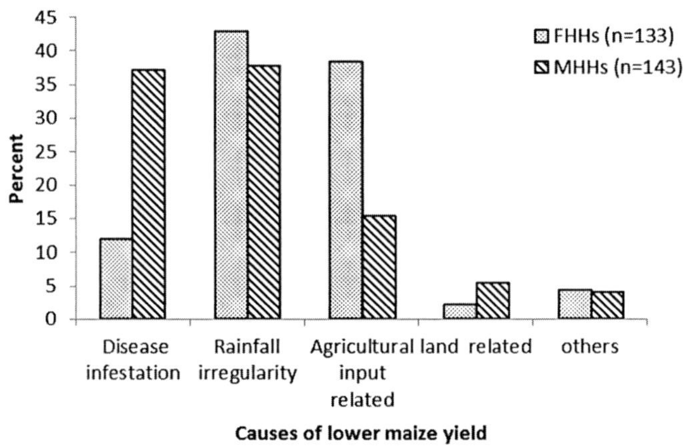

Source: Household survey, March to April 2011

Low agricultural productivity leads to loss of income for farm households. Yet, it is quite an irony that farmers complain of being unable to pay their debts as a result of the loan[7] they got from the government for the purchase of fertilizer and improved seed of maize. Under extreme situations, some households were even forced to sell out their livestock and pay back their debts. Thus, it is possible to observe some sort of vicious downward spiral leading to growing vulnerability and food insecurity. Both men and women FGD participants expressed their fear that the 'hunger season' in the year 2011 could be longer than in earlier years as they struggle to sustain their household food needs.

[7] A total of 520 *birr* (20 Euro) buys 6 kilos of maize, the amount needed for one *timad* of land. A full package of 50 kg of fertilizer and improved seed cost 1800 birr in 2011.

Low income induced by low agricultural productivity coupled with high prices of non-consumable commodities is another livelihood challenge for farmers in the study area. With the existing low purchasing power, the high price of soap, oil and gas has prevented poor women farmers from accessing basic necessities. "*Production is small and the money that we get from the market hardly buys us anything.*"

Economic hardship due to price instability is strongly felt even by those who have been given the title of "model farmers".[8]

"We don't sell our grain at fair prices. What we see is the price of grain decreasing while the price of processed goods, transport and non-consumable products is steadily skyrocketing. Regardless of this, all we hear in the media is about the price of grain increasing. The problem of rising prices for non-food commodities is rarely an issue for officials both at local and national level. Most farmers do not raise this as an issue at meetings because of lack of awareness. Yet, this is the major bottleneck even for the so-called 'model farmers' like me. "

A model farmer in Dobo Tuto Kebele

Livestock disease

According to the survey findings, nearly 80 percent of female-headed and 78 percent of male-headed households reported problems in raising livestock. Livestock disease is reported by 43 percent of female-headed and 47 percent of male-headed households. Nearly 26 percent and 27 percent of respondents in FHHs and MHHs respectively identified disease, together with others factors such as shortage of water and fodder (Table 9.2).

Table 9.2 Constraints of livestock rearing by type of household

Constraints of livestock rearing	**Type of household**					
	FHHs		**MHHs**		**Total**	
	№	**%**	**№**	**%**	**№**	**%**
Shortage of fodder	25	18.5	26	17.6	51	18.0
Shortage of water	7	5.2	5	3.4	12	4.2
Disease	23	17.0	32	21.6	55	19.4
Lack of fodder + water + disease	13	9.6	16	10.8	29	10.2
Shortage of water + disease	11	8.1	8	5.4	19	6.7
Shortage of water + fodder	18	13.3	14	9.5	32	11.3
Shortage of water + fodder + disease + knowledge	11	8.1	14	9.5	25	8.8
No problem faced	27	20.0	33	22.3	60	21.2
Total	135	100	148	100	283	100

∑ Livestock disease in FHHs=58 (43 percent), MHHs=70 (47.2 percent)

Source: Household survey, March to April 2011

In a mixed farming system such as the one in the study area, crop production and livestock rearing are highly interlinked. The reduction in crop yields is usually a threat for the produc-

[8] An honour given to economically best performing farmers within *kebeles*. They are assumed to be role models for other farmers in their *kebeles*.

tion of livestock since it also hampers natural forage[9] production. Poorly fed animals in turn are a problem during the planting season, since the planting season requires strong oxen for ploughing. Thus, this further limits crop production in a sort of vicious cycle. Livestock with poor access to food and water are also highly susceptible to disease. Diseases such as, *aba gorba* (blackleg) and a*ba sanga* (anthrax) are common threats to cattle in the study area. Infectious epidemic diseases such as *fengil* constrains poultry production, particularly where modern equipment is used, and female-headed households are the most affected since they tend to be attracted to poultry production.

"We advise farmers to keep the chickens inside their cage, however for some reason, they don't want to listen to us. They let them free. That's why they usually catch diseases such as fengil. And once they are infected there is very little chance that the chickens survive and the disease transmits from one to the other very quickly."

A DA in Dobo Tuto *kebele*

The expert blames the farmers for their negligence and misperception of the value of chickens.[10] Yet, the problem is also exacerbated by the shortage of medical personnel and veterinary services in the study *kebeles*. Key informant farmers revealed that at the time of data collection, there was only one health expert serving three *kebeles*. Weak administration at *kebele* level makes the provision of medical services extremely slow and it is difficult to meet the demand of farmers. For instance, in Dobo Tuto *kebele*, requests for medicine were turned down because the *kebele* authority refused to open new packages until 250 cases of infected chickens were registered in the *kebele*. The respondents in the other *kebeles* revealed similar administrative issues.

As discussed above, livelihood shocks hardly stand by themselves given "the interdependent pillars of livelihoods" in the study *kebele*. Crop failure, illness and the death of domestic animals are common livelihood challenges for both female- and male-headed households that are linked to one another. Yet, in FHHs the combined effect of these common shocks with other gender related disadvantages[11] seems to have a more negative impact on their overall productivity, weakening their ability to withstand these shocks at the household level. The following section aims to verify this statement through an investigation of traditional coping mechanisms and coping strategies used by female- and male-headed households.

9.2 Traditional coping mechanisms and coping strategies

The survey result revealed that households devise and employ a range of coping strategies in addition to the well-established traditional coping mechanisms so as to reduce damage and maintain their well-being amidst livelihood shocks and stresses. Based on the survey findings presented in Table 9.3 the traditional coping mechanisms that are widely used at the household level include various consumption arrangement methods, and various informal insurance strategies (helping hands, inter-household transfers and loans). In addition to these, livestock exchange, shifts in cropping decisions and income diversification through off-farm activities are substantially undertaken as important coping strategies in all of the study *kebeles*.

[9] Usually made out of crop residues

[10] Poultry is seen as a low-value economic undertaking.

[11] These are discussed in chapters 7 and 8. They include resource related constraints such as shortage of land, poor land preparation during the planting season, such as delay in ploughing, less use of agricultural inputs and other institutional barriers.

Table 9.3 Traditional coping mechanisms and coping strategies employed by respondent households

Types of coping		Type of household					
		FHHs		MHHs		Total	
		№	%	№	%	№	%
Traditional coping mechanisms	Consumption adjustment	11	6.3	8	4.5	19	5.4
	Informal Insurance	20	11.4	7	4.0	27	7.7
Coping strategies	Selling/exchanging livestock	10	5.7	26	14.7	36	10.2
	Shift in cropping decisions	32	18.3	12	6.8	44	12.5
	Income diversification through off-farm activities	12	6.9	8	4.5	20	5.7
Combination of traditional coping mechanisms and coping strategies		90	51.4	116	65.5	206	58.5
Total		175	100	177	100	352	100

∑coping strategies other than traditional coping mechanisms in FHHs=144 (82.3 percent), MHHs=162 (92 percent)

Source: Household survey, March to April 2011

A large proportion of FHHs and MHHs combine both traditional coping mechanisms and coping strategies in order to survive food crises and maintain resilience at household level. This applies to 51 percent of FHHs and 66 percent of MHHs. Coping strategies such as exchange of livestock, shift in cropping decisions and off-farm activities are employed by a large proportion of both female- and male-headed households accounting to 82 percent and 92 percent respectively. In what follows, each of the coping mechanisms and strategies are presented in detail placing a particular emphasis on the gender and wealth aspects.[12]

9.2.1 Consumption adjustment mechanisms

Similar to the "sequencing approach" emphasized in Dessalegn (1991) as well as Webb and Von Braun (1994), poor farmers start coping with food shortages by shifting or modifying their consumption habits four to five months prior to the main "hunger season". Likewise, among the study population, the reduction of meals from three times a day to one or two times per day becomes more common when households approach the "hunger season." As food stress increases, households tend to intensify these strategies by minimizing meals to save food for the next days. Moreover, consumption of lower-value grains such as maize, instead of a relatively pricy food item such as *teff* in wealthier households, as well as consuming *zengada* and poor quality *enset* in poorer ones are common mechanisms for overcoming the 'hunger season' in poor households regardless of the gender of household head.

In relatively well-off households whose production can sustain domestic consumption for six to eight months, reduction in the frequency of consumption is employed only during the rainy seasons where food stress is relatively high. These farmers are also active in the market three

[12] For discussion on different levels of wealth categories refer to Table 6.3

or four months prior to the onset of the main "hunger season".[13] Thus, they are able to ensure regular consumption by selling high-value crops and purchasing low-value grain or by selling sheep and goats and purchasing grain when grain prices fall. For instance, households sell *teff* and purchase *zengada*, maize or beans to ensure the continuous provision of food for household members.

Generally, the consumption adjustment mechanisms followed by households heavily depends on commodity exchange on the market. And this is regardless of the gender of the household head. Yet, resource-poor FHHs are more constrained in their involvement in the market as compared to labour-abundant and financially well-off farm households which are mostly MHHs. This is discussed further in section 9.2.5.

9.2.2 Informal insurance mechanisms

Informal insurance mechanisms in the study area function through membership or participation in village-based saving and credit associations, community-based organizations (CBOs)[14] and other forms of labour- and resource-sharing arrangements. These institutional mechanisms facilitate the exchange of labour, borrowing grain or money, bartering food for assets, setting up village- or community-level credit schemes, as well as the giving of psychological support and advice during illness and death or other adversities. The section below presents the findings of the study regarding some of the most well-known forms of informal insurance in the study area by giving emphasis to their relevance in terms of coping with shocks and stresses in each household group. Thus, the first section deals with the role of *ekub* and *idir*, followed by other resource- and labour-sharing arrangements based on the data gathered in the field.

A. *Ekub* and *idir*

In addition to formal credit services, informal credit and saving has a long institutional history in the study *kebeles*. Among the study communities, the tradition of saving for certain expenses, both in kind (livestock and grains) and in cash is commonly practised in all of the study *kebeles*. These take the form of small savings and credit groups such as *ekub* (a rotating saving and credit association), and funeral associations like *idir* (see also 8.2.4).

Regardless of gender, the survey finding shows that equally large numbers of MHHs and FHHs have membership in *idir* (Figure 9.3). But while *ekub* groups were once very popular, they are becoming less relevant to the study population. Only 3 and 4 percent of FHHs and MHHs respectively have membership in *ekub*. Similarly, membership in both *idir* and *ekub* hardly exists.

---

[13] Normally, the dry season is ideal for purchasing maize to be saved for the next rainy season.

[14] "Membership organizations made up of groups of individuals who have joined together to further their own interests and /or the interests of others" (Fischer 2006, 3).

Figure 9.3 Membership in informal social associations by type of household

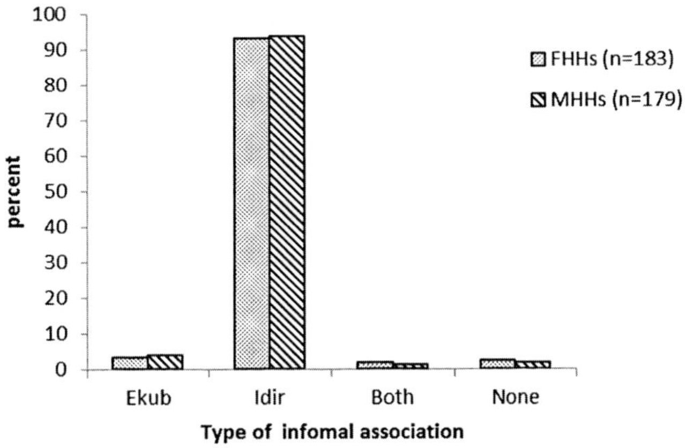

Source: Household survey, March to April 2011

Ekub

The qualitative information gathered in the study *kebeles* showed that *ekub* are primarily women's associations and can be formed based on residential proximity, or may be religion-based or may simply include women who are interested in benefiting from the savings and credit service. Savings through *ekub* are often very small and mostly the money is used to cover one-time expenses during holidays and feasts instead of long-term investments. For instance, during the FGD discussion with FHHs in Semen Shershera, respondents mentioned an *ekub* where a group of Muslim women were involved in saving a small amount of money on a weekly basis. The savings are mainly meant to cover costs during holidays and feasts. Yet, the study shows that *ekub* is dying out. There are various reasons for its gradual disappearance in the study area. Women FGD participants mentioned the growing access to formal credit services and the deteriorating economic condition of farm households and their resulting inability to save money as the primary reasons.

Even though the presence of *ekub* groups signalizes the existence of a local saving culture, expert opinion from Gurage zone micro finance office casts doubt on the efficiency of the households' decisions in relation to their pattern of expenditure. Thus, it is emphasized that the relatively large amounts of money spent on religious, cultural and social events such as marriage and religious events weigh heavily on the livelihood of households. In fact, the damage might even go to the extent of pushing households into situations of greater food insecurity. Government policies strongly assert the need for developing a culture of savings and loan repayment among rural communities. Even though this is meant in the context of formal credit institutions and savings and credit cooperatives, this need is also met by informal saving associations such as *ekub*. Yet, the system requires an attitudinal change on the part of the actors themselves.

Idir

Idir groups are community-based funeral organizations. Their primary role is to help the family of a deceased person by providing labour, money, food and emotional support. In the study *kebeles*, the support lasts for 8-10 days after the funeral takes place. The FGDs held with female- and male-headed households revealed that *idir* functions by substituting the role of family members during burial ceremonies. In the words of one FGD respondent: "*Idir means a brother, it is guest friendly and welcoming*." In the words of a male elderly participant: *"If I die, the first to be there to pick up my body is the idir. Not even my children. Children may probably be heard on the telephone."* In general, the responses indicate that *idir* has a very important place in the social life of the study population.

Idir groups in the study *kebeles* have their own organizational and administrative structures and are run by locally appointed elders, mostly men who are known to have a good social profile. In all the study *kebeles* there is no woman who holds such a position. The organization has its own written rules and regulations which its members are required to abide by. These consist of regulations regarding membership requirements, monthly contributions and restrictions and procedures of penalty when members act against the law. Depending on the nature of the crime committed, defiance of the rules might result in a penalty that ranges from paying a fine in cash to loss of membership status. However, some of the restrictive rules related to prohibition against engaging in any form of work on funeral days and the days following burial ceremonies are gradually being abandoned. This is primarily due to the frequent economic hardships and the increasing income needs of households to survive those hardships.

Lately, *idir* groups have broadened their scope and have started to play an important role in the socio-economic and political life of the study population. Serious illness, economic distress and family breakdowns and disputes are addressed using *Idirs* as a channel of communication. *Idir* are used as a medium in mobilizing resources among the community in case of emergency needs, acting as a buffer against shocks. *Idir* also control crimes and unlawful actions by community members, such as deforestation, misuse of public resources, and acts of theft. Therefore, i*dir* groups work in close collaboration with the police in combating crimes. They are also used to communicate political messages, government policies, and programmes initiated by the government to the community. Members are informed about current affairs and development programmes that are being implemented in the study area. For instance, the implementation of government fertilizer programmes is facilitated by using i*dir* as a medium of communication. Negotiations with the *kebele* administration use i*dir* as a primary channel for communication. Thus, for the study population membership in i*dir* is considered obligatory.

Discussions held with groups of women from all of the study *kebeles* revealed that the information most women have about *idir* does not go beyond its customary role of facilitating funeral ceremonies. According to discussions with women FGD participants, it was clear that most women are not aware of the multiple tasks of *idir,* particularly regarding its role in the on-going development programmes of the government. Moreover, in all of the study areas, the involvement of women in i*dir* is not more than being a member. There are no women members of committees or in administrative positions. The reason for this, according to *idir* committee members, is the absence of educated women in their community. Moreover, culturally, women in general are assumed to be less qualified and competent to handle complex social affairs at community level. Even though i*dir* serves the community regardless of sex, the absence of women in leadership positions within such organizations is a reflection of the relatively low status of women as compared to men. Information is the key to empowering women and enabling them to exercise leadership positions. More importantly, in the current situation where the political role of *idir* is increasing, it is critical that the needs of women are included in all decision-making processes. This is imperative to solve the multiple livelihood challenges faced by women farmers in the study *kebeles*. Yet, their absence in positions of power undermines their ability to benefit from development interventions, exacerbates their vulnerability and weak coping capacity and leads to the continuation of their low social status in their community. This is particularly damaging when proactive livelihood decisions are needed to cope with the changing socio-economic and environmental reality.

B. Labour- and resource-sharing mechanisms

Shortage of labour is a common problem in both FHHS and MHHs. But it is more severe among FHHs and the demand grows during times of livelihood crisis. In the study *kebeles*, labour-sharing arrangements are age-old practices that symbolize community-level cooperation in the face of shortage of labour and resources. According to the survey results shown in Figure 9.4, only 22 percent of FHHs and 37 percent of MHHs reported no labour shortage problems. The rest use various labour-sharing mechanisms. A disproportionate proportion of female respondents responded that they resort to asking for help when suffering from labour shortage. The finding shows 41 percent of FHHs against 10 percent of MHHs. Hiring labour at times of labour shortage is more of an option in MHHs than in FHHs.

Here, it is also important to note that the small proportion of women who identified sharecropping and hiring labour as a means of accessing labour does not contradict the findings given in chapter eight which showed a large segment of FHHs as benefiting from sharecropping. In this regard, two major explanations can be offered. First, it might be due to the fact that women do perceive share cropping as an integral part of their livelihood by which they mobilize not only labour but also a variety of productive resources such as oxen and managerial skills. Second, shortage of labour in the study area is mostly associated with labour-sharing arrangements such as *debo* rather than sharecropping, which is directly linked to the lack of a man to plough the land.

Figure 9.4 Type of labour-sharing mechanisms by type of household

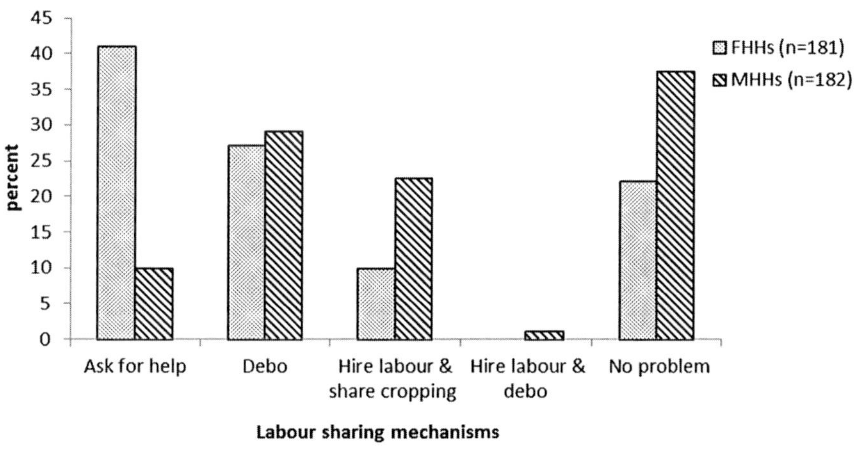

Source: Household survey, March to April 2011

Debo, a village-based labour sharing tradition, is employed by 27 percent and 29 percent of FHHs and MHHs respondents respectively. In the study *kebeles, debo* working groups are organized to relieve the work load of households during times when labour is most needed to undertake agricultural tasks such as harvesting and weeding, as well as labour-demanding social events. This kind of village-level cooperation also helps to reconstruct the economy of highly affected households following difficult crises. Traditionally, the arrangement takes the form of mutual assistance through labour exchange or food and does not involve cash payments. However, with increasing monetization of labour in the study *kebeles*, the rules and norms of these institutions are undergoing some transformations. The discussion with female and male respondents revealed that people are becoming less interested in tying themselves to working the whole day for food and drink because there are more attractive chances to work for money as a wage labourer in the towns. As a result, people ask for payment when undertaking such jobs. The gradual deterioration of labour-sharing arrangements such as *debo* in the study area negatively affects the poorest segment of the community, such as FHHs who need manpower before and after the planting season.

The survey shows that a greater percentage of FHHs as compared to MHHs turn to relatives and neighbours for free assistance, which is an indication of their weaker economic position as compared to their male counterparts. However, the dependence of women on such free assistance compromises their opportunity to make decisions in certain areas of their livelihood. For instance their ability to apply agricultural inputs such as fertilizers and improved seeds is constrained as it requires tasks which they can manage only with the help of relatives and neighbours.

The survey finding in Figure 9.5 shows that nearly 83 percent and 67 percent of FHHs and MHHs respectively face a shortage of oxen to undertake agricultural activities. Village-level resource-sharing mechanisms are also employed to mobilize other resources such as oxen. Among these, sharecropping, *mekenajo (*a resource sharing mechanism to acquire additional oxen for ploughing) and borrowing are the most widely used practices in the study area. *Mekenajo* is an age-old popular form of resource-sharing practised by farmers with only one ox. Traditional plough farming requires a pair of oxen, and the purpose of *mekenajo* is to join with another farmer who has only one ox, so that both can plough with a pair. The survey finding in this regard shows a larger proportion of MHHs (38 percent) who use this arrangement, as against 22 percent of FHHs. FHHs tend to depend more on sharecropping. This accounts for nearly 30 percent of FHHs, compared to only 7 percent of MHHs. Nearly 31 percent and 22 percent of FHHs and MHHs respectively depend on relationships and networks to borrow oxen.

Figure 9.5 Types of oxen-sharing arrangements by type of household

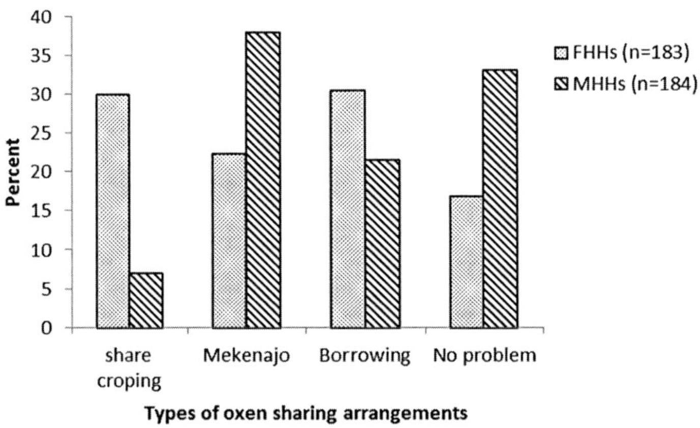

Source: Household survey, March to April 2011

From the above discussion of informal insurance mechanisms, it is clear that social ties are potential sources of coping mechanisms against shortage of resources. FGDs with female- and male-headed households revealed that the former largely depend on borrowing or free assistance to mobilize resources at times of stress. MHHs, on the other hand, are more likely to depend on other mechanisms to access resources instead of dependence on reciprocal ties. This has some implications with regard to the decision-making power of women farmers. Dependence on free assistance forces women to maintain good relations with their relatives. This requires them to socialize by throwing feasts that are costly. Moreover, they are expected to show submissiveness while negotiating and bargaining during sharecropping deals. It is worth noting that informal insurance mechanisms function better with idiosyncratic shocks than with common shocks. In the case of the latter, people's will and capacity to provide free assistance to others is more restricted, since resources will be stretched to meet many household demands (see section 7.1 and section 9.3 on the weakening of mutual and reciprocal social support systems).[15] This all compounds to show women farmers' have weaker coping mechanisms as compared to their male counterparts.

9.2.3 Selling and exchanging livestock

In his case study of the Wollo famine, Dessalegn documented variations of the pattern of asset disposal among households. The process is generally selective and gradual. Households start by selling their smaller stock, followed by young cattle, then cows and finally oxen (Dessalegn 1991).

The findings with regard to the pattern of asset disposal in the study area revealed both similarities and differences with respect to these findings. Both men and women FGD discussants unanimously agreed that selling cattle such as oxen is considered the last option to cope with a crisis. But they stressed that decisions no longer follow a strict sequence, depending on the type and level of severity of the crisis faced by farm households. For instance, at times of heavy income stress, households often resort to selling sheep and goats so as to get money for the purchase of fertilizer (see 8.2.5). Such strategies are quite unusual and depending on the severity of income shortage the items that are sold can even range from small ruminants such as sheep and goats to cattle. To know whether households sell their livestock as a means of coping with livelihood shocks or crises, the respondents were first asked to identify the type of livestock they sold in the last five years. Table 9.4. shows the type of livestock households sold in the last five years.

[15] Dercon (2000) argues that in situations where a large number of households are commonly affected by a certain shock, there is a minimum chance for coping through community transfers.

Table 9.4 Types of livestock sold during the last five years

Type of livestock sold in the last five years	Type of household					
	FHHs		MHHs		Total	
	№	%	№	%	№	%
Oxen	18	10.3	39	22.0	57	16.2
Cows	5	2.9	8	4.5	13	3.7
Donkeys	2	1.1	-	-	2	0.6
Sheep+goats	14	8.0	10	5.6	24	6.8
Poultry	7	4.0	6	3.4	13	3.7
Cattle*+small ruminants*	25	14.4	21	11.9	46	13.1
Did not sell any	103	59.2	93	52.5	196	55.8
Total	174	100	177	100	351	100

*cattle include oxen and cows, small ruminants include sheep and goats

Source: Household survey, March to April 2011

The survey result in Table 9.4 shows that a greater proportion of FHHs (59 percent) as compared to MHHs (52 percent) did not sell any livestock in the last five years. The proportion of MHHs who sold oxen is double the proportion of FHHs. Fourteen percent of FHHs and 12 percent of MHHs sold both cattle and small ruminants. Combining those who sold only oxen or cows and those who sold both cattle and small ruminants in the five-year period, 28 percent of FHHs sold cattle or small ruminants as compared to 38 percent in MHHs.

Yet, these figures above do not tell the reason behind the asset disposal; in other words, whether it was done as a short term strategy to fill income or food gaps or as a long term strategy to cope with unusual shock or stress or for the purpose of investment. Therefore, an attempt was made to gather information regarding the reasons for selling the livestock. The survey results summarized in Table 9.5 help to clarify this further.

Table 9.5 Reasons for the sale of livestock

Reason for selling livestock	Type of household					
	FHHs		MHHs		Total	
	№	%	№	%	№	%
To purchase grain	19	10.9	18	9.8	37	10.4
To buy other livestock	25	14.4	35	19.1	42	11.8
To pay debts	13	7.5	12	6.6	25	7.0
To pay tax	10	5.7	6	3.3	16	4.5
To construct a house	1	0.6	7	3.8	8	2.2
To lease land	-	-	2	1.1	2	0.6
Did not sell any	103	59.2	93	50.8	196	54.9
Others	3	1.7	10	5.5	13	3.6
Total	174	100	183	100	357	100

Source: Household survey, March to April 2011

The largest number of both female- and male-headed households mentioned purchasing other livestock, followed by purchasing grain. Through FGD discussions, it was found out that depending on the season and the relative prices of grain and livestock, households largely use exchange of livestock on the market to cope with food gaps during the "hunger period." Households can exchange oxen for oxen, oxen for cows or cattle for small ruminants. Exchanging of cattle for sheep and goats is becoming an important coping strategy since the cost of fodder and lack of water has put constraints on the raising of cattle in the study *kebeles*. Livestock can be used as a form of exchange for other animals or it can be sold as a source of income to fulfil other livelihood demands. However, the traditional gender role that assigns the task of selling cattle and small ruminants to men makes it difficult for women in FHHs to actively involve themselves in the process of exchange in the market. They mostly depend on relatives and close friends to act for them. This means that they are unable to take timely measures when shocks and stresses occur that affect the provision of food and other necessities in their households. This coincides with that of the study findings of Tröger *et al.* (2011) with respect to women in the Nyangatom society.

Generally, from the survey finding it can be seen that exchange of livestock on the market is one form of coping strategy against shocks in both types of household, yet female-headed households do so under certain gender-based constraints.

9.2.4 Shift in planting decisions

Confronted with a shorter rainy season and the risk attached to growing maize, households in the study area are forced to re-think their cropping pattern. For instance, assigning a greater part of their land to low-return but also low-risk traditional crop varieties and leguminous crops, and planting trees for market[16] are some of the widely used practices that are used to minimize the risks of crop failure. For instance, in addition to maize, farmers assign their land to crops like *zengada* and legumes such as peas and haricot beans which are better suited to sandy soil and grow well despite the increasingly short rainy season. Growing of fruits in addition to perennial crops is also used as a coping strategy against income shortage.

The data obtained from the survey show that nearly 41 percent of FHHs as compared to 27 percent of MHHs planted one type of crop during the year 2010. Maize is the major crop, grown by 30 percent of FHHs and 18 percent of MHHs. The survey finding shows an extremely diverse response regarding households who produced more than one type of crop in that particular year (see Appendix V). To show if there is consistency with the qualitative information about the growing of *zengada* and legumes, attempts are made to summarize the number of times these crops appeared in the responses given by female- and male-headed households. The results showed that FHHs are less likely to diversify the crops they grow. Growing more than one type of crop requires additional tasks which FHHs, with their limited labour, time and agricultural skills, are less likely to manage.

During the focus group discussions held with MHHs and FHHs in Semen Shershera *kebele*, the respondents unanimously agreed on the relative profitability of growing fruits and vegetables such as avocado, cabbage and potatoes, but also coffee and chat, particularly during the dry seasons. Yet, shortage of water makes such practices difficult in all of the study *kebeles*. The few households who are able to benefit are those whose land has the potential to retain water during the rainy season, or the few with access to water-harvesting holes.

[16] In Dirama and Semen Shershera *kebeles*, growing eucalyptus is increasingly becoming a lucrative business, due to the high demand for young wood in the market.

FGD participants emphasized that a year with a good harvest is usually followed by poor rains and a sharp deterioration in food supplies in the subsequent year. As a result, assigning part of the farm land for growing economically profitable trees, such as eucalyptus, is becoming popular, in addition to other income-generating activities. The drastic increase in domestic demand in relation to young wood and charcoal has facilitated such land-use changes. Planting trees is mostly practised by young couples for reasons of better profit and manageability as compared to growing crops.

Few households are able to apply their knowledge to overcome the effects of shortage of water by switching to certain well known but less practiced plant varieties in the study area. A model farmer in Dirama *kebele* told of his success with growing *enset* (false banana) and sweet potatoes[17] and stressed his surprise that many households in his village do not do the same. *Enset* is a highly drought-resistant plant with a good yield potential. Almaz and Anke (2004) have verified the contribution of this plant to the food security of households. Yet, not many farmers grow *enset* in the study *kebeles*. Expert interviews revealed that this is frequently due to reasons such as threats from wild animals, diseases and lack of knowhow. Moreover, processing the plant for consumption is traditionally a women's task and most MHHs are sceptical about it, since it is assumed it only benefits women.

9.2.5 Income diversification through off-farm activities

Participants of FGDs in all of the study *kebeles* emphasized that the increasing uncertainty in the timing of rainfall has not only disrupted the local agricultural calendar, but also eroded their reliance on agriculture *per se* as a sole source of livelihood. Respondents stressed that intensifying livelihood diversification outside of agriculture is gradually becoming a necessity to survive unprecedented periods of crisis. Working as daily labourers in towns like Butajira, and and taking on petty trading are common among the study population. The reality of the growing trend of diversification into non-agricultural income-generating activities was sarcastically expressed by an elderly key informant in Dirama *kebele*, who said: "*Gone are the old good days, where everybody was a farmer.*" The man's association of being a farmer with the 'old good days' is an indication of his regret regarding the gradual erosion of the norm and the culture that he grew up with, in which one's identity depended on being a farmer. The respondent saw it as a threat to local culture that such age-old practices are undergoing transformation. This is also a clear evidence of the gradually weakening traditional coping mechanisms in the study area.

The survey revealed that around 95 percent of FHHs and 90 percent of MHHs were engaged in one or more than one type of off-farm income-generating activity at the time of data collection. Yet, qualitative inquiries through a FGD with female- and male-headed households revealed that these activities are not uniformly available throughout the year in the study area. The results of the discussion are summarized in Table 9.6.

[17] A low-risk crop

Table 9.6 Seasonality of off-farm activities

Season	Off-farm activities		
	FHHs	MHHs	
		Young couples	Elderly couples
Dry/ *bega* season	Petty trading	Men engage in daily labour, construction jobs. Women engage in petty trading	Less likely to take part in off-farm activities
Wet/ *kiremt* season	Slack season for petty trading	Men engage in wage labour	Less likely to take part in off-farm activities

Source: Focus Group discussions, February 2011

Women are generally more active in petty trading during the dry season than in the wet season (*kiremt*). Even though the wet season is the main "hunger season", in which consumption gaps are severe, poor connecting roads, less customers and poor market infrastructures constrain women's mobility during this period of the year.

Male-headed households are divided into younger and elderly age groups in terms of their participation in non-agricultural activities. Most young couples have little land thus often take up daily labour activities so as to augment their income. Jobs such as being a security guard in the district town of Butajira, engaging in paid agricultural field work or construction work as a daily wage labourer in towns are becoming very common off-farm ventures for this category. According to the respondents, there are two possible reasons why relatively elderly couples do not engage in off-farm activities. The first reason is their relatively better wealth status and the second is their age, which means they are no longer so active and mobile.

Table 9.7 presents the findings from focus group discussion with regard to the various types of off-farm activities carried out in female and male-headed households in each of the study *kebeles*.

Table 9.7 *Kebele*-level distribution of off-farm activities

Type of household	Type of off-farm activity by kebele			
	Beresa	**Dobo Tuto**	**Dirama**	**Semen Shershera**
FHHs	Petty trading of grain	Petty trading of grain and wage labour	Petty trading of grain, selling firewood	Petty trading and selling firewood
MHHs	Men: Wage labour inside and outside the *kebele*, working as security guard Women: petty trading	Men: Wage labour inside and outside the *kebele*, roof building, digging, cutting wood or trees Women: petty trading	Men: Wage labour inside and outside the *kebele* Women: petty trading	Men: Wage labour both inside and outside the *kebele*, cutting trees and transporting loads Women: petty trading

Source: Focus group discussions, February 2011

In the context of diversification into off-farm activities, the following section shows how women and men actors understand and interpret the social rules and norms that guide their everyday interactions through the decisions and actions they make while undertaking these activities. It also attempts to show how resource availability in male- and female-headed households influences these decisions. It also identifies gender-related constraining factors in off-farm activities.

Off-farm diversification in FHHs

The summary of responses obtained from women FGD participants in Table 9.7 shows that women both in female- and male-headed households are short-distance petty traders. Petty trading primarily involves selling grain and firewood. As petty traders women involve themselves in a sophisticated arrangement of exchange of food and non-food items with the primary aim of filling food gaps and other household necessities. Women in Beresa *kebele* who are involved in petty trading purchase pepper from Enseno[18] market on Thursdays and Saturdays and sell it on Tuesdays and Fridays in markets in Butajira[19]. In other cases they buy grain from Jole[20] market and sell it at Hamus Gebeya[21]. A woman from Semen Shershera explained: *"I buy 11 tasa[22]maize for 20 birr (a little less than 1 €) in Jole market and sell only 10 tasa of the grain for the same price at Hamus Gebeya. This way I am able to secure 1 tasa of maize for dinner."* Local markets, namely Jole, Hamus Gebeya, Enseno and Butajira, are the common destination for women petty traders. Mostly, women walk to Butajira markets on foot or use a donkey if they possess one. Figure 9.6 shows pictures of the local Friday Butajira market captured during the field work.

[18] Located around 17 km from the district town of Butajira
[19] Located around 5 km from the study *kebeles*
[20] Located around 5 km from the district town of Butajira
[21] Located around 15 km from the district town of Butajira
[22] One *tasa* is equivalent to nearly 0.7 kg.

Figure 9.6 Friday Butajira market

Source: Field work, January to March 2011

Women face various barriers while carrying out this type of petty trading activity. First, due to the relatively great distance of markets from their villages and lack of affordable transportation women are restricted to short-distance mobility and small-scale trading of perishable and low-value products. Second, the lack of affordable communication services limits women's knowledge about market conditions. FGD discussants revealed that they depend on oral communication to exchange information about market prices. This can happen either on their way to the markets or while they are in the markets. The negative effects of the lack of reliable means of communication are particularly felt during the rainy seasons when market prices are relatively unstable. In this regard, due to the absence of market information and fewer clients, female heads are usually unable to stay active in the market. Given the fact that the rainy season is the main "hunger season" in the study area, the negative impacts of the weakening of off-farm activities on coping ability are more strongly felt by FHHs at this particular time than at any other time of the year. In addition to all these challenges, time-consuming and laborious domestic tasks, such as hauling water, do not allow women to spend much time in the market, thus minimizing the benefits they obtain from their petty trading activities.

As intensification of off-farm activities becomes a necessity for survival, women are also increasingly becoming attracted to the wage labour market. Yet, women in general and female heads in particular face certain gender-based institutional barriers. Most jobs outside agriculture are restricted to men with some level of education. Those women who are able to take such jobs are generally paid less than their male colleagues. At the time of data collection, the wage for men was between 15 and 20 *birr*, while women were paid 12-15 *birr* per day. The main reason behind such wage gaps, according to male FGD respondents is that any activities undertaken by women are considered less laborious than what men normally do. This corresponds to the findings of Quisumbing and Yohannes (2005) who examined the determinants of participation in, days worked, wages, and earnings from off-farm activities using household survey in rural Ethiopia.

Off-farm diversification of activities in MHHs

Due to the gender division of labour, responsibility for provision of food and devising survival strategies for the family in MHHs is mostly assigned to women. Similar to women in FHHs, women in MHHs contribute to the household economy through their active involvement in petty trading. As one elderly man in Dirama *kebele* put it, "*The women and their donkeys do much of the work and carry the heavy burden of the household economy, they are the engines of the economy.*" A male FGD participant in Semen Shershera *kebele* boldly expressed the enormous contribution of women, saying, "*All we can say is we are ruled by our women.*" A DA in Semen Shershera *kebele* further stressed the economic contribution of women in the *kebele*, saying: "*Women's leadership is not only in female-headed households, even in the so-called male-headed households only* a *few men really lead their household. The government programme targets women in the kebele, not because there are no men. It is because the source of income for households is small and limited to petty trade in which women's management skills are generally better. When women take up the management of such tasks, results are fast and can be seen easily. Women are honest managers when it comes to their household affairs.*"

These responses generally express the enormous contribution of women in supporting the household economy by generating income and providing for the needs of household members.

Men in poorer MHHs also engage in off-farm activities, usually wage labour. Relatively wealthy male farmers are less likely to work as wage labourers. Instead, they engage in long-distance exchange of agricultural commodities such as grain, livestock[23] etc. In the case of wage labour, depending on the availability of jobs in their village or in towns, men seek employment outside farming throughout the year. They usually take up jobs such as roof-building, digging, cutting wood or trees, security guard, engaging in construction works, sewage and pipeline works, or digging wells in towns like Butajira. The growing construction works in towns, for flourishing NGOs and government offices, are potential sources of jobs outside agriculture. Yet, male farmers still complain about the seasonality of jobs, low wages, shortage of available jobs close to their villages, and long working hours as some of the major problems in the wage labour market.
A young male farmer in his 30s said:

"I don't have land. I usually plough for women farmers and sometimes rent land and grow crops for my household. I sell fodder for farmers in addition to working as a wage labourer in Butajira town. Gradually, my income grew and I was able to buy land close to Butajira and build my own house. I am now engaged in long-distance grain trading and my income is growing. I see a better future than ever before."

Generally, three important points summarize the findings presented so far. First, in the study area, working outside agriculture in the form of petty trading and wage labour is an important coping strategy for the majority of FHHs and MHHs in the poor category. The finding of the study in this regard agrees with the findings of Dercon and Krishnan (1996)[24] in Tanzania and

---

[23] Travelling long distances means selling high value and large volume products to wholesalers and cooperatives, which usually translates into higher income. On the other hand, travelling frequently over short distances usually targets local spot markets where goods are sold directly to consumers in smaller volume.

[24] They found that poor farmers typically enter into activities with low entry costs: firewood collection, charcoal, collecting dung cakes, casual agricultural wage employment, etc. Entry into high-return non-crop activities, such as cattle rearing or shop keeping, is restricted to richer households with access to capital. Non-agricultural wage employment is generally restricted to those with some education.

Ethiopia. Second, the findings show that MHHs enjoy more options to take up more than two types of off-farm activities due to the better flexibility of time and labour availability at household level. Third, the survey makes it clear that for MHHs the availability of resources, and guiding norms in the form of gender roles, can facilitate decisions in respect of coping with livelihood shocks, as well as for the betterment of livelihoods. In contrast, female heads are trapped in their gender-based customary roles which affect their decision to engage in off-farm activities. Women are mostly active in low-return petty trading activities, or perform low-paid "women's' jobs" in the wage labour market. This is shown above by two responses obtained from the field: the "hand to mouth" market-based coping strategy employed by a woman farmer versus the "wealth accumulation" strategy of a male farmer.

9.3 Weakening traditional coping mechanisms: households' response to income stress

Taking into consideration the study findings on traditional coping mechanisms presented so far, this section is intended to investigate which kind of coping options are available to farm households at times of income stress, as well as the decisions female- and male-headed households make under such circumstances. The aim is to shed light on the degree of vulnerability of the households at times when other well-being needs appear. Accordingly, households were asked to identify the strategy they would use if they had to earn 100 *birr* within a period of 7 days. As shown in Table 9.8, compared to MHHs (36 percent), a larger proportion of FHHs (nearly 59 percent) do not have any concrete means of obtaining 100 birr in such a short time. The self-insurance method based on savings is used more often in MHHs in which the figure showed 16 percent, as against 6 percent in FHHs. Instead, FHHs were found to be more likely to sell livestock, with 10 percent of FHHs as compared to 5 percent of MHHs.

Table 9.8 Options that would be used to get 100 *birr* in an emergency situation by the type of household

How to get 100 *birr*		**Type of household**					
		FHHs		**MHHs**		**Total**	
		№	**%**	**№**	**%**	**№**	**%**
Sell livestock		18	10.3	9	5.2	27	7.8
Selling grain		16	9.2	25	14.4	41	11.8
Sell firewood		1	0.6	2	1.1	3	0.9
Sell private property		5	2.9	7	4.0	12	3.4
Informal insurance mechanisms	Ekub	2	1.1	-	-	2	0.6
	Idir	1	0.6	-	-	1	0.3
	Borrowing from relatives	15	8.6	35	20.1	50	14.4
	Ask friends & relatives for a gift	2	1.1	1	0.6	3	0.9
Own savings		11	6.3	28	16.0	39	11.2
Wage labour		-	-	4	2.3	4	1.1
Can't get it		103	59.2	63	36.2	166	47.7
Total		174	100	174	100	348	100

∑Informal insurance mechanisms in FHHs=20 (11.5 percent), MHHs= 36 (20.7 percent)

Source: Household survey, March to April 2011

Informal insurance mechanisms are identified by 12 percent of FHHs as compared to 21 percent in MHHs. Borrowing from relatives ranks first in this category accounting to 9 percent and 20 percent of female- and male-headed households respectively. Here, it is important to note that the relatively small proportion of FHHs in this category does not contradict the findings given in section 9.2.2[25]. In this regard, it should be underlined that monetary assistance and assistance in the form of labour or oxen can have different meanings for the study population in which women have more chance to receive the latter. FGDs with FHHs in the study area revealed that women refrain from borrowing money from relatives and friends since it requires quick repayment, and failing to pay back the money on time might damage their social reputation. Moreover, private money lenders give attention to the creditworthiness of the borrower before giving credits. For instance, oxen and land possession, as well as the condition of the house, are usually considered in order to judge whether an individual is creditworthy or not. Unfortunately, access to credit through private loan providers is hardly an option in FHHs.

In general, ranking the responses presented in the table above, borrowing from friends and relatives, withdrawing one's own savings and selling assets such as grain and livestock are important emergency response mechanisms in MHHs, in decreasing order. In the case of FHHs, the order is selling assets, informal insurance and own savings.

The high proportion of FHHs who failed to any of the options provided, despite financial stress, is a clear indication of the fading away of traditional coping mechanisms such as informal social networks. The findings in the study area gain support in the study of Tröger *et*

[25] Section 9.2.2 shows relatively better figures with regard to the involvement of FHHs in various labour sharing and oxen sharing arrangements.

al. (2011, 32) in which the gradual weakening of traditional institutions that are based on mutual thrust and reciprocity led to what can be described as a growing trend of a "more individualized way of life". Such situation presents a higher degree of vulnerability for already economically poor groups such as FHHs. It leads to a situation in which poverty persists with poor capacity to access food and low level of other well-being indicators, such as health, education, housing, etc. In addition to this, the fact that a greater proportion of FHHs depend on selling liquid assets such as livestock shows the narrow options available for FHHs as compared to MHHs. For the already resource-constrained households, removing assets like livestock can have a long-term consequence for their future coping ability at times of severe food gaps. This is reflected in Corbett (1988), Devereux (2001), Watts (1983) and Dessalegn (1991).

9.4 Interim summary

The chapter started off with the presumption that female-headed households have weaker coping strategies against livelihood shocks and stresses as compared to their male counter parts.

The study shows that illness of a family member, death of domestic animals, crop failure and resulting shortage of income are the most serious forms of livelihood shocks experienced by both types of households. Traditional coping mechanisms such as various consumption adjustments, and various informal insurance strategies (helping hands, inter-household transfers and loans) as well as coping strategies such as exchange of livestock, shifts in cropping decisions and off-farm livelihood diversification strategies are employed by both female- and male-headed households. Fifty one percent of female- and 66 percent male-headed households, combine the traditional coping mechanism with coping strategies. Coping strategies such as exchange of livestock, shift in cropping decisions and off-farm activities are employed by a large proportion of both female- and male-headed households accounting to 82 percent and 92 percent respectively.

Consumption adjustment as one form of traditional coping mechanism is carried out regardless of the gender of household head. Commodity exchange at the market plays an important role in the process of consumption adjustment, but resource-poor FHHs are more likely to involve the market as compared to the labour-abundant and financially well-off MHHs. Informal insurance mechanisms identified in the study *kebeles* include membership of or participation in area-based saving and credit associations, community-based organizations (CBOs) and other forms of labour- and resource-sharing arrangements. Regardless of the gender of the household head, a large proportion of the survey respondents are members of *idir,* while membership in *ekub* is rare in both female- and male-headed households. Even though traditional labour-sharing arrangements such as *debo* are gradually being abandoned and undergoing certain transformations in the study area, they are still important sources of coping mechanisms against labour stress in both female- and male-headed households, with relatively higher involvement of MHHs than FHHs. A relatively large proportion of FHHs as compared to their male counterparts depend on relatives' free assistance at times of labour stress. Oxen-sharing arrangements, including sharecropping, *mekenajo* and borrowing, are widely used village-level cooperation schemes. Yet, around 38 percent of MHHs depend on *mekenajo* arrangements compared to only 22 percent of FHHs. In this regard, a greater number of FHHs either depend on free assistance in the form of borrowing or enter into sharecropping arrangements to access oxen during the planting season. The study also found out that reciprocity is an important factor in free informal transfers, which ties women with certain social obligations. It is also less reliable at times when common shocks affect the wider population.

In the case of livestock exchange on the market as a strategy of coping with shocks, the survey found out that a greater proportion of FHHs (59 percent) as compared to MHHs (52 percent) did not sell livestock in the last five years. Among those who reported selling livestock, a large proportion of both female- and male-headed households mentioned purchase of grain and livestock as the reason. Yet, given the fact that livestock markets are considered a "male domain" in the study area, women in FHHs face institutional barriers to take timely measures at times of shock and stress. This affects the provision of food and other well-being necessities at household level.

With regard to shifts in planting decisions as a coping strategy against shocks, the survey result generally shows that FHHs are less likely to grow more than one type of crop as compared to MHHs. The survey showed that 41 percent of FHHs as compared to 27 percent of MHHs planted one type of crop during the year 2010. Growing more than one type of crop requires additional tasks which FHHs, with their limited labour, time and agricultural skills, are less likely to manage. With regard to households' engagement in off-farm activities, the survey revealed that around 95 percent of FHHs and 90 percent of MHHs were engaged in one or more than one type of off-farm income-generating venture during the year of data collection. In general, women are short-distance petty traders of grain and firewood, while both petty trading and wage labour are important income-generating activities for MHHs. The study identified various constraints to engaging in off-farm activities in both female- and male-headed households. Yet, women face greater institutional barriers. In this case the study found out that women are engaged in activities that conform to traditional gender roles which restrict their ability to benefit from off-farm ventures. Moreover, their limited resource base restricts them to small-scale meagre income-generating activities.

The study shows that a large proportion of FHHs (59 percent) as compared to 36 percent of MHHs do not have a response mechanism to emergency financial need showing the gradual weakening of traditional coping mechanisms such as informal social ties. Among the category of those who are able to resort to such mechanisms, a large proportion of FHHs sell assets, while only 20 percent and 8.6 percent of MHHs and FHHs are able to borrow from friends and relatives. Given their low liquid asset possession of FHHs, removing assets for short term financial needs can carry further livelihood risks with long-term consequences for their future coping ability at times of severe food gaps.

Generally, in the face of certain unprecedented environmental, economic and social events which are referred as shocks and stresses, both female- and male-headed households actively make coping decisions by engaging in what Giddens call "a reflexive-discursive dialog". These decisions are made in the face of a range of competing societal rules and norms from which actors must choose following their understanding and interpretation of their everyday life circumstances. Such process leads to action, in a form of coping strategies which are employed by men and women actors to survive shocks and stresses and ensure food security. Yet, the extent to which actors are able to engage in such processes at a discursive level is influenced by the amount of power they are able to wield which is linked to their access to resources. In the institutional setting of the study area, where unequal gender power relations are highly manifested in the prevailing unequal access to resources between men and women farmers, the coping strategies employed by FHHs meet severe constraints. As a result, coping strategies simply serve to fill short-term food demands on a daily basis. Yet, they fail to support livelihoods and ensure long-term food security. Therefore, the findings in this chapter support the presumption that FHHs have weaker coping strategies than MHHs.

10 External intervention

This chapter begins by presenting the findings of the study regarding the major types of food security programmes (FSPs) that are run in the study *kebeles*. This is followed by an assessment of the general merits of these programmes in terms of improving the food security of households in the study *kebeles*. Its contribution in terms of narrowing the gender gap in vulnerability and food insecurity is examined in the third section. Finally, an attempt is made to assess the empowerment efforts that are being undertaken in the study area, thereby showing the existing gaps. This is believed to indicate key entry points for further action.

10.1 Food security programme in the study *kebeles*

In the study *kebeles,* the food security programme components under implementation include the direct support, the public works/ community asset building and the Household Asset Building Programme/ family package programmes. These are carried out under the Productive Safety Net Programme. For instance under the public works scheme, activities such as building and school renovation, building of terrace on hillsides, construction of roads, building seedling centres, prevention of flood damage by building dams or conversion of barren and vulnerable lands into cultivable crop lands were carried out in the study *kebeles*. Table 10.1 shows the proportion of beneficiaries of the public works and direct support programmes of the PSNP in four of the study *kebeles* disaggregated along gender line. Nearly equal proportion of female- and male-headed households are beneficiaries of public works. The direct support program targets a larger proportion of female-headed households as compared to male-headed households.

Table 10.1 The number of beneficiaries of PSNP in the study *kebeles*

Kebele	**Public works beneficiaries from public works**						**Beneficiaries of Direct support**					
	FHHs			**MHHs**			**FHHs**			**MHHs**		
	№	**%**	**T***	**№**	**%**	**T***	**№**	**%**	**T***	**№**	**%**	**T***
Beresa	69	36	190	209	42	503	11	6	190	5	1	503
Dobo Tuto	46	61	76	108	31	346	16	21	76	3	0.8	376
Dirama	78	41	189	173	47	371	18	10	189	16	4	371
Semen Shershera	71	45	157	204	55	371	18	11	157	7	2	371
Total	264	43.1	612	694	43.6	1591	63	10.3	612	31	1.9	1591

*Total number of households in each *kebele*

Source: Hand-out prepared by Meskan district 2011

The household asset building programme/ the family package programme focuses on providing credits for households who are interested in agricultural or non-agricultural income-generating activities. For instance, in all of the study *kebeles* women, particularly in physically and economically weak FHHs, are provided with poultry production packages. Sheep and goat breeding, livestock fattening and marketing are other areas where farmers can obtain credits[1] to benefit from the existing family packages. Moreover, youth packages are made available for unemployed young men who are less than 30 years old, in order to help them engage in non-agricultural ventures as a means of income generation. Sand production in Dirama *kebele* is one such example. There are similar initiatives in Semen Shershara *kebele*

[1] Farmers are expected to return the loan within a 2-year period.

even though certain financial constraints were mentioned by a development agent as a barrier to running the programme at the time of data collection.

10.2 Merits of the FSP in the study *kebeles*

According to the information gathered through FGDs and key-informant interviews, some of the merits of the FSP are related to saving lives and preventing asset depletion, as well as perception, behavioural and attitudinal aspects.

***Saving lives*:** An interview held with the head of the health extension office in Dirama *kebele* revealed that among those whose health is the most compromised at times of food shortage are children, FHHs and women in MHHs with weak support from their spouses. Development transfers in the last couple of years have enormously assisted these vulnerable groups to fill the consumption gap and avoid extreme hunger, thus reducing mortality related to food shortage and malnutrition. The interview also revealed that around 25 percent of households within the safety net programme are able to avoid diseases related to food shortage. The findings of the FGDs with men and women farmers, and official documents of the district food security office also confirm some improvement in this regard. A non-beneficiary male FGD participant explained: *"If a community member is able to afford his child the necessary health treatment, he may save his child's life. Indirectly we save the time spent on visits, giving condolences and funeral ceremonies, and we save the money our idir spends on helping the family of the deceased. Thus, indirectly the benefit is both for the beneficiaries and the non-beneficiaries."*

Prevented asset depletion: A key informant development agents in Beresa, Dobo Tuto and Dirama *kebeles* revealed that the PSNP has prevented the depletion of liquid assets such as livestock and land. Consistent information was obtained from focus group discussions and secondary sources at the district level, regarding the success stories of women farmers who have been able to build assets by purchasing oxen or small ruminants, and of households that have been able to change from grass roofs to corrugated iron sheet. There are cases of women who have lost their land through land contracts[2] as a result of food shortage and who were able to get their land back through the PSNP. Moreover, women participants in the FGDs frequently stressed that the food and cash transfers of the PSNP had to some extent relieved their income stress with regard to paying school fees, health costs and other social obligations which they otherwise manage through selling livestock or contracting land. A DA in Semen Shershera *kebele* said: *"Well-disciplined and workaholic households are the ones who are able to exploit the benefits of the programme. In this case, women farmers are good role models. They are very cautious in planning and managing their household. They plan how to spend their money and they are the ones who sacrifice the most to send their children to school."*

***Perception, behavioural and attitudinal aspects*:** The findings from the FGDs show that there is a positive feeling among beneficiary and non-beneficiary farmers towards the public works dimension of the PSNP. The various environmental rehabilitation and infrastructural development works, and the achievements they have witnessed in their *kebeles* have taught farmers the difference they can make through cooperation. The introduction of a group-managed water pump scheme in one of the study *kebeles*, Beresa, has increased the confidence of the farmers and shaped their thinking, as they gradually learn the benefits of applying technology to enhance their production and manage their own livelihood on a sustainable basis.

[2] Respondents said that there have been cases of poor female-headed households renting their land for as little as 20 or 30 birr.

From a gender perspective, female FGD participants mentioned that the economic relief as a result of government interventions has to some extent reduced crimes such as theft in the study *kebeles.* As a result, participants unanimously agreed that in addition to its contribution in terms of filling food gaps, the introduction of the PSNP has brought them greater social security as compared to earlier times. In addition, the accessibility of schools which are now closer to the study *kebeles* has increased the chance of girls enrolling in school. Increased access to education can reduce HTPs such as early marriage and other forms of violence against women.

At household level, FHHs have made a positive impression among agricultural experts due to their active engagement in the FSP. According to the opinion of a male DA, "*women farmers have showed their will, loyalty and strong commitment to work with the food security task force at kebele level. For instance, they are always at the forefront during public works, oftentimes they are the first to pay advance payments for improved seeds and fertilizers, and are the first in returning government debts.*" Women beneficiaries have in turn welcomed the recognition they are given in various food security programmes. In this regard, there were some instances where the family package programme primarily targeted women farmers. For instance, in Dobo Tuto *kebele,* during the year of data collection, out of the 12 households that the family package programme selected for a grant of 4000 *birr* for the purchase of oxen, eight of the beneficiaries were FHHs.

10.3 Assessment of the FSP: low graduation rate / high dependence on transfers

The primary aim of the food security programme is to enable chronically food insecure households stabilize their asset base and gradually allow them accumulate assets. As a result, it is assumed that they become food sufficient first, then sustainably food secure. This gradually leads to their independence from the PSNP, and then from the FSP. In other words, graduation[3] out of the programme is the ultimate objective of the Food Security Programme (MoARD 2009b). Unfortunately, by the time of data collection, there was no full impact evaluation study[4], report or official document showing the progress of households during the course of the implementation period of the PSNP. However, key informant interviews with *kebele* level officials revealed that the graduation rate in all of the study *kebeles* is very low.

In this regard, information obtained from Beresa *kebele* showed that out of 17 candidates proposed by the *kebele* food security taskforce, only 6-8 people have been able to graduate since the start of the programme and none of them are female-headed households. In Dobo Tuto *kebele*, the *kebele* was planning to graduate 7 people who have been able to improve their houses from straw thatched *tukuls* to corrugated iron sheet, and to increase the number of their oxen from one to two or three. In Dirama, *kebele,* only a few (no specific number is mentioned) were ready to graduate in the year of data collection. Moreover, at the district level, an interview with a food security expert revealed that there was no graduate in the year 2002, despite the plan to graduate 5000 households in 24 beneficiary *kebeles.* There is very limited information about the progress of beneficiary households towards graduation. Moreover, the existing data on the wealth status or progress of beneficiary households in all of the study *kebeles* is fragmented and incomplete.[5] Generally, with the existing data, it can be roughly

[3] A term used to show a level where a beneficiary no longer receives outside assistance.

[4] An evaluation report obtained from Meskan District Agriculture and Development Office (2011) states that the task force at the district level did not spend enough time with the *kebele* level task force to work on the issues of dependency and graduation.

[5] An interview with an expert in Semen Shershera revealed that households who are able to send their children back to school after dropping out, who are able to increase their asset possession through purchase or exchange of cattle (from goats and sheep to cows and from calves to oxen), and who are able to improve the condition of

said that the progress towards graduation of beneficiary households in general and FHHs in particular is quite low in all of the study *kebeles.*

Using primary data gathered from in-depth interviews with extension workers, officials, beneficiaries and non-beneficiary households and the information obtained through official reports, the following section attempts to highlight the gaps observed in the implementation phase of the FSP. The aim is to illuminate those areas where the programme fails to achieve its objective for a large number of farm households in general and FHHs in particular. In the meantime, the section sheds light on the underlying reasons for the low success of the programme in bringing FHHs in to graduation.

Experts' versus local farmers' opinion

Several factors have caused delay in the success of the FSP in the study *kebeles*. These can be detected at district, *kebele* and household levels among beneficiary and non-beneficiary households. According to the 2011 evaluation report obtained from Meskan district, the misguided perceptions of higher officers and experts, both at district and *kebele* level, as well as of beneficiary households, constitute one of the pitfalls observed during the implementation process. In the case of the former, the document states that the PSNP is misunderstood as a form of aid rather than as an input for development works. The evaluation documents say: *"There is a misconception among beneficiaries and non-beneficiaries of the program to the extent that transfers made are taken as a monthly salary. For many farmers, being a beneficiary has a positive connotation since it is associated with getting recognition from the government. Thus, those who are not beneficiaries feel discriminated; they take it as if the government hardly knows them or as if they are denied their constitutional right."*
Similar opinions were gathered through FGDs and in-depth interviews with beneficiary and non-beneficiary households. There were mentions of beneficiary farmers who have gone as far as selling their assets, removing their assets to other *kebeles*, or destroying their houses and crops on their farm land, so as to give the *kebele* officials the impression that they are poor. A key informant explained: *"On the one hand, those who remain poor year after year boast around saying that they are still on the beneficiary list. Those who have shown improvement on the other hand perceive their improvement as something undesirable, even as a threat, for they might get deleted from the beneficiary list. Graduating out of the programme is given a negative connotation."*

Nevertheless, further in-depth interviews with beneficiary households revealed gradual shifts in the perception of farmers, while illuminating the existing perception differences between farmers and the programme-implementing body. A beneficiary in Dirama *kebele* expressed his opinion by saying: *"How long can someone depend on begging to sustain himself? Instead, a couple of enset in one's own yard can be a sustainable dinner. We regret that the nature of the rains and the soil condition could not help us to do this."* Contrary to the 'dependency syndrome' emphasized above, this farmer's response shows a preference for self-reliance, as it is assumed to lead to sustainability on a long-term basis. In this case, the existing dependency on external transfers is due to the insecurity of livelihoods caused by environmental factors.

In the context of this study, it was a challenging task to adequately capture the issues mentioned above in their full complexity. From the responses given above it can be argued that

their houses are considered as the ones approaching the graduation phase. Yet, the existing data is only based on informal observations.

the confusion among the beneficiary and non-beneficiary households is primarily due to misconceptions regarding the term "graduation" itself. Clearly, the implementing body needs to clarify such terms in a way that avoids negative connotations for both better performing and poorly performing households.

In addition to the above perception-related constraints, focus group discussions with beneficiary farmers and key informant DAs identified both the rigid nature of the package programme and reaction of farmers as major constraints in the implementation of the programme. For instance, in-depth interviews with DAs in the study *kebeles* revealed cases in which the loan provided to farmers under the family package program was fully or partially used for purposes other than the one fixed in the package programme. Some had used it for petty trade, while others used it to purchase donkeys or cell phones, send their children to school, build houses, or other purposes. The money transfers were used by some to pay back loans taken in advance from private money lenders. In extreme cases, heavy debts in MHHs led to family breakdowns. Similar information was also given by focus group discussants in the study *kebeles*. In this regard, the findings are in line with Tröger *et al.* 2011 in which it is said: "*The safety net program has changed the farmers' coping mechanisms in a negative way... safety net works only inefficiently at the moment because livestock productivity is close to zero and credits for buying livestock are unproductive.*"

10.4 Assessing the components of the programme through a gender lens

Gender-based analysis of the FSP requires assessment of the components of the intervention by taking into consideration the asset and capabilities of both gender groups. In order to investigate gender biases with regard to implementation of the FSP, the following section discusses the family package and the public works components of the FSP as two examples.

Family package: access to modern poultry production and other technologies

During the FGD discussion with women farmers, women discussants mentioned their preference for traditional poultry keeping methods, which are less time consuming and less risky than the modern poultry production methods introduced by the government. Women perceive modern poultry production as an undertaking that requires commitment and care. As a result, among able-bodied as well as older and physically weaker female heads[6], it is considered a cumbersome, time- and labour-demanding activity. For destitute FHHs it is also considered a financially difficult undertaking, since it requires properly built cages which they are less likely to be able to afford. These households also refuse to take the package for fear of default.[7]

Livelihood improvement strategies offered to the local communities should, therefore, take into consideration the capabilities of the different types of households that they are designed to serve. And for such interventions to succeed, gender should be mainstreamed into the policies and programmes. The participants pointed out that implementation of this kind of package programme should be based on proper consultation with both men and women. The appropriate components can then be offered, based on each household's need and interest. Imposing inappropriate components on poor households is destined to failure, or only partial success, since it aggravates indebtedness and fails to respond to the demands of men and women farmers in need.

[6] The poultry package programme most often targets elderly female-headed households, since it is considered an easy undertaking.

[7] Lack of skills and health risks are mentioned as the major constraints associated with poultry production. This is also discussed in chapter nine under livelihood shocks.

Public works dimension of the PSNP

The public works dimension of the PSNP is designed to facilitate the building of public assets in the study *kebeles*. Easy and constant access to road infrastructure and communication services, schools and markets close by, or supply of clean water and sources of energy play an enormous role in alleviating the time and labour stress felt by FHHs. Yet, in the year of data collection, all of the study *kebeles* had uniformly poor access to water, energy and road infrastructures. The slow progress in this regard is partly due to the low attention given to follow up, quality control and evaluation of the public works performed under the food-for-work scheme. As a result, the overriding problems of women in general, and FHHs in particular, such as poor access to clean water supply and sources of energy, are not addressed by the programme. Given women's additional workload from domestic chores and productive activities, it can be argued that the public works scheme within the food security programme has overburdened women, while hardly solving their problems of shortage of labour and time.

10.5 Local empowerment processes and policy gaps

The empowerment of women has been the leading development approach pursued by the government of Ethiopia in the process of poverty reduction and in the effort to achieve sustainable development goals over the last two decades (see 3.2). In this section, an attempt is made to assess the existing empowerment efforts in the study area, based on the information obtained from the field, followed by a discussion of the existing gaps.

Access to and control over land

Increasing rural women's access to land is one of the areas in which government policy has made major strides. The findings of the study on access to land in chapter eight showed that nearly all the surveyed FHHs have access to agricultural land registered in their name. Most women have positively welcomed land certification which guarantees their right to use their land. Having a user right over a plot of farm land has psychologically boosted women's confidence and allowed them to exercise their agency better than ever before. As one FGD respondent said about the benefits of having ownership of land: *".... nowadays I can say, 'no you can't sell the ox'. I can say that my man is spending our money inappropriately and take this case to the appropriate body. I know that I have the right to say this and I will do it. I now have that ability (capacity), so I will do it. Previously women did not have economic ability to do so. So they were highly exploited by their husbands."* This response shows that improved access to resources like land has created a favourable environment for women to start exercising the rights the constitution grants them.

Yet, the findings obtained through FGDs, key informant interviews and participant observation do not lead to the conclusion that women are adequately empowered to be able to make decisions at various level. Female head FGD participants from Beresa *kebele* raised various issues surrounding the control and utilization of land which constrained their potential to exercise livelihood choices in their favour. One of the participants explained: *"One year the sharecropper does his job very well and the other year the next one might be too careless. He might plough the land at a time when others plant seeds and might overlook other related activities. If my son is around, he will follow up and give the sharecropper some feedback. But I have to always be very friendly with the sharecropper. He may even give up and refuse to give further assistance if he feels offended by the feedback. Our life is just a struggle!"*

The above response and the findings presented in 8.2.1 show that despite the policy provisions that grant land ownership to women in the form of user rights, women still face various constraints that prevent them from utilizing and fully benefiting from their land. This can be partly explained by women's non-involvement in ploughing, and their weak bargaining position during sharecropping negotiations, as the above example demonstrates. As a result, it can be argued that the government's effort to empower women through provision of access to land has failed to ensure women adequate control over and benefit from their land.

Awareness of women's rights and gender equality

Creating an awareness of women's rights and gender equality, with the aim of eliminating stereotypes, discriminatory perceptions and practices, is another dimension of the empowerment process in the study area. FGD-based inquiries in this regard showed that both men and women in all of the study *kebeles* have had some information about women's rights and gender equality. However, of all the *kebeles,* women in Beresa *kebele* had a unique exposure to such information, due to the extensive engagement of a women's empowerment NGO called Progynist. The impact of the role played by this kind of NGO with regard to awareness creation is observed in various areas of women's life. A participant in Beresa *kebele* explained: *"Before Progynist entered our kebele, there were many problems. For example a woman had no power to make decisions concerning her property, and her entire movements were controlled by her husband and men in the village. Now I know that I can go out on a field and plough my land, I am not afraid that people will condemn me because of culture. But I do not have the time and the resources to do so, so I give my land to a sharecropper."*

The gender advocacy work by the above-mentioned NGO has also enhanced the awareness of households with regard to the negative consequences of harmful traditional practices and gender-based violence. It has encouraged women to challenge norms that promote such acts. Women who are aware of their rights gradually become sceptical towards the customary judiciary system and opt to seek justice through the formal court. The findings presented in section 7.3 show that over the years, the number of cases handled through the formal court system have increased in all of the study *kebeles*, the highest number being in Beresa. This shows that increasing access to information in the study *kebeles* has enabled women to start making proactive decisions in various areas of their livelihoods. Yet, while the existing situation shows the growing agency of women, it also raises important questions regarding the role of the traditional judiciary system in the empowerment process.

The findings of this study show that the legacy left in the study area by NGOs such as Progynist is enormous in terms of empowering women. By the year of data collection, any activity of NGOs (including Progynist) in areas related to women's right was non-existent in all of the study *kebeles*.[8] However, the existing situation is still far from the elimination of stereotypes, discriminatory perceptions and practices which are the underlying structural problems working against the equality of men and women in the study *kebeles.* A key informant gender expert at the district office said "The *women themselves are not changed yet. Many women who are victims of violence do not come looking for legal assistance; they remain silent because of fear of stigma and exclusion by their community members. In other cases, even those women who decided to take their cases to the police and district courts face barriers such as lack of adequate resources to follow their cases. As a result, they withdraw from the formal justice system and go back to mediation by the shimagiles."*

This response indicates that despite the efforts to create awareness of women's rights, the empowerment process has not yet brought substantial changes in the practical life of women. Institutional barriers such as cultural norms and an unfavourable judiciary environment, as

[8] This is partly due to the "NGO law" passed in 2009, known as the "Charities and Societies Proclamation", which restricts the activities of NGOs working on women's rights and democracy issues, among others.

well as resource-related barriers, still prevent women from exercising their rights. All these combine to undermine the success of empowerment efforts in the study *kebeles*.

Organizational structure

At the time of data collection, the offices working on gender issues at district level were the rural women's department, accountable to the department of crop production at the district agricultural office, and the women, youth and children department at the district administration bureau. In addition to these, three gender experts were working voluntarily on a part-time basis at the district level, with no workplace and no monthly salary. At *kebele* level, there were no women's affairs departments responsible for mainstreaming gender (see Appendix VI). In this regard, the only structure functional at *kebele* level is the women's Saving and Credit Institution, yet with no political mandate. Investigation through key informant interviews with gender experts at the district revealed that the organizational structures for women's affairs departments exist only on paper and are hardly functional even at district and *kebele* levels. Neither were there gender focal groups that entertain complaints and provide legal support for women.

The women's affairs departments at the district agricultural office face severe constraints in their organization and have not succeeded in ensuring gender mainstreaming in their different activities. Besides, since they have an equal position with other departments they cannot directly influence the plans and activities of other departments. Lack of adequate provision of the necessary human, financial and material resources has been the major hindrance for the proper functioning of the gender offices both at *kebele* and district level.

The current development programme Growth and Transformation Plan (GTP) emphasizes strengthening women's associations and organizations as an important tool to facilitate gender equality. Information obtained through field experience indicated the presence of women's credit and saving associations in all of the study *kebeles.* Yet, all of these associations work only in the economic sphere and hardly play any role in the social, political and cultural life of women. Moreover, they do not have any link to the local council or cabinet at *kebele* level.

Generally, even though efforts have been made to build the capacity of gender offices and personnel in the past few years, the organizational structures for implementing gender policies were found to be fragmented both at the district and *kebele* level. In view of these gaps, the existing situation both at district and *kebele* level is far from realizing what is envisaged in the policy document.

10.6 Interim summary

Various components of the national food security programme have been implemented in the study area with the aim of assisting households to fill their food gaps, prevent asset depletion and build up assets both at household and community level. These include direct support, food for work or the household asset building programme (HABP), components of the PSNP, and the family package programme under the FSP.

This study has identified not only a decrease in mortality and prevention of asset depletion, but also perception, behavioural and attitudinal transformations of people as the positive achievements of the food security programme (FSP). The study also shows certain general and gender-specific challenges that have hindered the effective implementation of the programme in the study *kebeles*. Lack of awareness and erroneous perceptions of the programme among farmers, and differences of perception between experts/officials and beneficiary farmers are identified as some of the barriers to the smooth running of the programme. These problems have led to unintended outcomes as beneficiary farmers act by re-adjusting the program to serve their own livelihood needs further affecting the coping strategies of farming households.

With regard to the gender dimension of the existing implementation gaps, the poultry project of the FSP did not consider the shortage of time and labour in FHHs, which led to their decision to stick to traditional poultry. Moreover, the public works dimension of the PSNP has hardly addressed issues of water, energy and road infrastructures which are the major challenges for rural women in general and FHHs in particular. In this regard, it can be asserted that the family package programme has overlooked the psychological willingness and motivation of female-headed households. Even though FHHs are targeted in various ways, their heavy involvement in the programme is not necessarily translated into meaningful benefits. Overall, the progress towards the graduation of beneficiary households in general and FHHs in particular is quite low in all of the study *kebeles*.

With regard to the major empowerment processes that have taken place in the study area over the past few decades, the findings in this chapter show that increased access to land has boosted the decision-making ability of women in general and FHHs in particular. Yet, the utilization and control over their land is hampered by women farmers' inability to plough their land. This minimizes the benefits they enjoy from their land. Progynist's intervention in Beresa *kebele* has played a key role in creating awareness of women's rights and gender equality among the study population until it ended its activities in 2009. Yet, the existing situation is still far from bringing changes in terms of eliminating stereotypes, discriminatory perceptions and practices against women in all of the study *kebeles*. With regard to the organizational structure of the women's policy implementing body, the study shows that gender offices are non- existent at *kebele* level and those at the district level are only present in fragmented state. Lack of personnel and financial resources, and inability to extend to grass root level, are identified as the major constraints.

11 Major findings and policy implications

In Meskan district, recurrent drought and environmental degradation pose a serious threat to the food security and coping strategies of rural people. Gender as a key social variable influences the vulnerability of women to such external shocks. Before I started this thesis, my prior research experience in the study area and my engagement in gender courses in the past shaped my thinking on gender issues. Bringing meaningful change to the disadvantages faced by Ethiopian women in general, and female-headed households in particular, requires enhancing the agency of women to enable them to challenge and question the institutions in which they are involved in their day-to-day life. These institutions promote and/or constrain their decisions and actions, and can undermine their well-being and status at the household and community levels, and beyond. This thesis hypothesized and confirmed true the presumption that in the face of common external livelihood shocks, unequal gender power relations frame the vulnerability context of rural households and are the prime contributing factor to food insecurity and weak coping strategies of FHHs as compared to their male counterparts. The study offers a unique insight into the complexity that gender components add to the challenges of poor farmers on the ground.

Using the sustainable livelihood framework and the "theory of structuration", the study shows how female- and male-headed households combine the resources at their disposal and translate them into different coping strategies to sustain their food security. The study raises important questions regarding the institutions that carry unequal gender power relations, the interplay between actors' agency and these institutions in the social field, how such interactions frame the vulnerability context of men and women, mediate access to resources, and influence the process of attaining food security and capacity to cope against shocks at household level. The following section discusses the major findings of this study in the context of the above-mentioned conceptual frameworks.

11.1 Major findings in light of the sustainable livelihood framework and the "theory of structuration"

The study set out with four major objectives, three of which were presented as hypothesises. (Refer to 1.3 and 1.4). Each of them is treated separately in four different chapters. In this section the major findings of the study are highlighted in terms of the theoretical frameworks used in this thesis.

(a) In the face of institutional pluralism, informal institutions shape the vulnerability context of female-headed households.

Rural people in the study *kebeles* have a substantial knowledge of the on-going social, environmental and economic transformations that are taking place in their surroundings. Unpredictability and variability of the *belg* rains, disruption in the local agricultural calendar, increasing shortage of labour and time to undertake domestic and productive chores are some of the situations shaping the external vulnerability of farm households. Yet, the ability of households to respond to the changes, which is the internal dimension of their vulnerability, is largely a function of the institutions that guide social relations.

In the study area, formal and informal institutions are at times competing and conflicting with each other. This thesis hypothesizes that in the face of such institutional pluralism, informal institutions shape the vulnerability context of female-headed households.
The study has revealed that three major institutions, namely traditional marriage and family, gender roles, and the customary judiciary system are the most dominant patriarchal institutions shaping the mode of life in the study area. A thorough investigation of the functioning of

these institutions has revealed that age-old and firmly established informal institutions carrying gender-biased norms, perceptions and customary practices govern social interaction and social relationships. The unequal gender power relations that are inherent in the rules, norms and practices that make up these institutions put women in a disadvantageous position where they become subject to discrimination, subordination and marginalization in all walks of societal life.

In certain areas where the influence of formal laws and the current economic down-turn is heavily felt, the existence of some of the well-established age-old institutional structures has been threatened. This has been followed by the partial transformation and replacement of customary practices such as polygamy and abduction. But in general, patriarchal structures persist as typical features of rural life in the study area. The behaviour and actions of both men and women farmers, shaped by processes of identity specification and internalization of their social position in society, reinforce gender-based stereotypic views and attitudes, ensuring the continuity of patriarchal institutions. By internalizing their subordination and inferiority to men, female heads carry with them the legacies of patriarchal structures, such as being a 'housewife' even in the absence of a male spouse. Similar to women in MHHs, female heads are trapped by their traditional gender roles, behave and act by conforming to the most gender-appropriate norms and codes of conduct.

Unequal gender power relations have clear implications for the livelihood of women farmers in the study area. For instance, women do not enjoy equal access to resources, and they exercise restricted agency during social interactions in which they have weak bargaining power. Livelihood decisions and actions which are based on such skewed power relations fail to effectively respond to the external shocks and stresses that threaten food security in the study area. Thus, the study has confirmed the hypothesis that in the face of institutional pluralism, informal institutions largely shape the vulnerability context of female-headed households.

(b) With less access to resources and decision-making, FHHs face greater food insecurity than their male counterparts.

The study results show that even though food insecurity is a common problem of both FHHs and MHHs in the study *kebeles,* FHHs face wider food gaps than MHHs. Resources such as land, oxen, labour and time, credit, agricultural technological inputs and water infrastructures are important livelihood resources determining households' food security in the study area. The study reveals that FHHs have smaller plots of land and enjoy less control over their land as compared to their male counterparts. They own fewer oxen, and are disadvantaged by lack of education, time, labour and finance, which prevents them from enjoying access to expert advice, agricultural inputs such as fertilizer, improved seeds and water for various purposes. Despite their involvement in saving and credit cooperatives, they engage in less lucrative and only small-scale income-generating activities, which minimizes the benefit they gain from the service. Most of these resources are acquired through social relationships that are played out in various institutional arenas, such as family, market, community and the state. Thus, less access to these resources in FHHs is a reflection of various factors that have their roots in the existing gender-based norms, values and customary practices that govern these relationships. For instance, widely held perceptions which designate women as "weak farmers" lead to smaller land and oxen possession by women and make them marginal players in the farming sector. In the face of institutional plurality, the customary judiciary system weakens women's capacity to enjoy full control over their land. Deeply entrenched gender roles put heavy pressure on time and labour in FHHs, constraining their access to agricultural technologies and inputs.

Resources play an important role in the livelihood of farm households by determining farmers' capacity to act in their social environment. Thus, improved access to resources enhances the quality of decision-making and choices made with regard to livelihood options. The findings in the study area show that the rigidity of gender roles and shortage of labour and time prevent FHHs from carrying out agricultural activities such as ploughing. They engage in sharecropping, which reduces the benefit they gain from their land. Due to their weak bargaining power in conflict situations, they exercise limited control over their land. In undertaking off-farm activities, they tend to engage in small-scale and low-return activities. Thus, poor access to resources in FHHs leads to decision-making and actions which are less effective in filling food gaps. Generally, the findings have proved the hypothesis that with less access to resources and poor decision-making powers, FHHs face greater food insecurity than their male counterparts

(c) As compared to their male counter parts, female-headed households have weaker coping strategies against livelihood shocks and stresses.

Interlinking livelihood shocks and stresses, such as illness of a family member, death of domestic animals, crop failure and shortage of income, all threaten the coping strategies of female- and male-headed households in the study area. Traditional coping mechanisms in the everyday practices of farm households are deeply intertwined with the routine livelihood strategies of rural people who continuously struggle to escape crisis after crisis to meet household food demands and other necessities. In this regard, consumption adjustment methods, and various kinds of informal insurance (helping hands, inter-household transfers and loans) are identified as common forms of coping mechanisms. The study found out that free informal transfers tie women through certain social obligations, as most of these arrangements involve reciprocity. Moreover, FHHs find themselves in a more vulnerable situation, as this kind of coping strategy is less reliable at times when common shocks affect the wider population. The study revealed that emergency financial needs cannot be met by a large majority of FHHs, showing the disintegration of traditional institutions as a reliable source of coping mechanisms. As a result, a large proportion of both female- and male-headed households depend on certain coping strategies to survive unusual events.

In their choice of coping strategies, FHHs in the study area face various constraints that largely have institutional roots. For instance, despite the huge dependence of FHHs on exchange of commodities on the market as a primary means of consumption smoothening, they get marginal benefits as they are only involved in small-scale and low-return petty trading activities. Moreover, in areas which are considered a "male domain" such as exchange of livestock on the market, they are unable to take timely measures by exchanging livestock at times of shocks and stresses. Generally, labour- and time-related stresses which are mainly rooted in their gender roles combine with lack of resources to undermine the benefits they gain from exchange on the market. It is hard for them to enter the labour market as the existing gender roles exclude women from engaging in activities that by tradition are considered "men's jobs." In this regard female heads are still trapped in their traditional gender roles which undermines their ability to benefit from the job market outside agriculture.

Generally, even though female heads of households just like their male counterparts are able to respond proactively to adjust to unusual environmental, economic and social transformations and the resulting livelihood shocks and stresses, their coping decisions are largely constrained by the deeply established rules and norms. Due to the unequal gender power relations that play out in these gender-biased institutions, the coping strategies employed by FHHs fail to provide all the household's needs, and they face severe challenges with regard to filling food gaps. By implication, the empirical evidence supports the hypothesis that FHHs pursue weaker coping strategies than MHHs.

(d) The role of food security programmes and gender policies in empowering women and addressing the gender gap in food security in Meskan district.

Both food insecurity and gender inequality are pervasive in the study area. Tackling these two dimensions of poverty has been the focus of external policy interventions in all of the study *kebeles*. For instance, all of the study *kebeles* are among those food insecurity prone *kebeles* in Meskan district that are targeted for government transfers through the Food Security Programme (FSP). Both direct support and the food-for-work as two major components of the PSNP, and the family package programme or the household asset building programme of the FSP have been implemented. The transfers aim to strengthen local coping strategies that are practised at household level by reconstructing the fragile environment, rebuilding households' economy, facilitating recovery from shocks and eventually enabling rural households attain a more sustainable livelihood. So as to narrow the gender gap in food security, the implementation of the FSP followed an approach of targeting women in general and FHHs in particular as its primary beneficiaries. The interventions had some positive impacts, such as decreasing mortality, or preventing poor households from depleting liquid assets such as livestock. It has also strengthened the culture of work through its food-for-work programme, unlike the previous aid-oriented approach. With regard to policies related to gender, the land distribution and certification that has been implemented in the study *kebeles* has improved rural women's access to land. Empowerment efforts such as creating awareness on women's rights, campaigns against HTPs and violence against women, and improving the accessibility of the formal judiciary system to women, has been carried out to some extent through government and an NGO intervention.

Despite the above stated efforts, from the investigation made in this study, the policies that are designed to tackle the multi-dimensional problems of FHHs have been affected by multiple constraints in their implementation phase. The FSP has simply focused on targeting women for projects that are considered to be in the "women's domain." Yet, female heads were unable to benefit fully from certain programmes. For instance, the family package programme has overlooked the psychological readiness, behaviour and motivation of FHHs, while the public works of the PSNP have hardly solved the problems of shortage of water, energy and road infrastructures which are the primary concerns of female heads. In the case of gender policies, the desired level of empowerment as envisaged in the National Women's policy has not been achieved. For instance eliminating stereotypes, discriminatory perceptions and practices against women in all of the study *kebeles* still remains a challenge. The bodies responsible for implementing women's policy lack adequate capacities, including personnel and financial resources. Most importantly, the absence of formal structures to implement policies meant to empower women at grassroots level weakens the link between the policies and the target population that they are meant to serve.

The lack of a horizontal relationship between the implementing organs of FSP and the gender policies lead to certain discrepancies. In this regard, the state machineries responsible for ensuring gender mainstreaming have failed to do so at grassroots level. A rough observation of the implementation process also reveals gaps related to the lack of participation of stakeholders, the local men and women farmers in the design and implementation of FSP. Farmers do not have the power to exercise their agency and make a clear impact on the intervention. As a result, it seems that implementation of the FSP in the study area follows a simple top-down approach which leads to unexpected outcomes other than the one intended. For instance, the re-adjustment of some of the programmes by the beneficiaries themselves is one manifestation of the lack of consultation of beneficiary households.

In the face of these gaps, the FSP and gender policies face stiff challenges with regard to attaining women's empowerment and narrowing the gender gap in food insecurity in the study area. The limited information obtained from the field showed that the progress towards graduation from the FSP programme among beneficiary households in general and FHHs in particular is quite low in all of the study *kebeles*.

11.2 Policy implications

The success of the current development strategies, policies and programmes in improving livelihoods and closing gender gaps in household food security in Meskan district should be assessed with a field-based, in-depth understanding of the lives of men and women farmers whom such efforts target. Based on the results obtained in this empirical study, the major policy implication that can be drawn is that policies should be geared towards improving the existing gender power relations. This would lead to greater well-being for women, which can translate into enhanced food security and effective coping capacity against shocks and stresses. Some of the policy implications of this study are presented below.

Women and the plough culture

The government of Ethiopia under the EPDRF regime has introduced a number of reforms that aim to improve the life of women in the country. Among its rural policies is the Rural land Administration and Land Use Proclamation No 456/2005. The proclamation guaranteed every rural household the right to hold and use rural land. Women farmers were able to ensure control over their land through the land certification programme which was implemented in many rural parts of the country, including the study *kebeles*. Since then, significant improvements have been observed in terms of ensuring security, avoiding disputes and narrowing gaps in conflict resolution which to a great extent victimized women farmers. However, in spite of such measures, women farmers are not still able to plough their land, and this limits the potential benefit they can gain from their land. There are several reasons for this, including the hectic and tedious household chores and their motherly responsibility of taking care of their children that hardly leaves them time to farm their plots of land. In addition, culturally constructed gender roles, patriarchal perceptions and social taboos discourage women from taking up activities such as ploughing, which is considered a man's job. These problems are slowly being tackled, but so far there has been no significant change in the lives of women farmers; they still continue to be victims of unfavourable economic and social power relations that alienate them from undertaking certain economic activities. There is obviously a need of cultural transformation. And one thing that would speed up the process would be introducing time-saving and easily applicable farm implements which can be used by women farmers.

Public infrastructures

As agriculture is threatened by environmental changes, off-farm activities such as petty trading are an important form of coping strategy to meet income and food demands in the study area. Even though petty trading is taken up by women both in female- and male-headed households, poor public infrastructures such as difficult access to potable water, bad roads, lack of means of transportation and lack of information and communication services are conditions that undermine the benefits to be gained from petty trading activities. For FHHs additional factors, such as shortage of labour, worsen the impact of these constraints, putting additional pressure on the time available for petty trading activities. Thus, provision of services such as water supply, road and communication infrastructures is something that should be given priority to enable women to engage in productive activities.

Institutional plurality in the justice system

Violence against women reinforces poverty and deprives women of their self-worth and confidence to realize their potential in the economic, social and political spheres. Harmful traditional practices and gender-based violence are common in the study area. In this regard, the customary mediation system of *shimagiles* has a long history of handling violence cases, as well as maintaining peace and harmony at community level. Even though this kind of arbitration system ensures long-term conflict resolution in contrast to the formal court of law, the study findings show that the traditional justice system has a number of flaws which disfavour women. In the course of conflict resolution much value is given to maintaining the existing patriarchal status quo and local culture. This comes at the cost of the well-being of women. In recent years, the district court and the police have been increasingly appealed to by women who prefer their case to be mediated by the formal court of law. Nevertheless, access to the court is difficult for poor female farmers. As a result, such cases are often handled by the customary court.

Following the CEDAW principle of the universality of rights which Ethiopia adopted in her constitution, states must not discriminate against female citizens under the pretext of culture and religion. Thus, it follows that the government is responsible for the well-being of all its citizens without discrimination. Yet, the recognition given to customary and religious laws can be seen as a barrier, since customary laws are not free from gender biases. Such institutional pluralism is a challenge to the proper functioning of the formal justice system and raises doubts about the effectiveness of the legislative provisions for eradicating gender-based violence. It is clear that legislative reform *per se* cannot bring gender equality without proper implementation in the field.

Under the current situation, where formal and informal judiciary systems co-exist, there is a need to make improvements in both systems in the light of the needs of poor rural women. In the case of the customary system, the government should consider facilitating community forums to openly discuss and reassess the role of *shimagiles* in handling violence. In the case of the formal system, serious consideration should be given to enhancing the capacity of the judges, reducing bureaucratic procedures and making the court more accessible to poor women. The efficiency of the judgment process has to be improved to avoid delays in the handling of cases of violence. A clear demarcation should also be set to differentiate between cases that should be handled by the customary court and those that should be brought to the formal court. A more efficient justice system would help to prevent violence from happening by challenging attitudes, beliefs and practices that promote the oppression of women.

Women's policy implementing organs at grassroots level

Women, to be able to use their legal rights, should be able to overcome the powerlessness that arises from their economic and social dependence at community level. In this regard, one of the empowerment approaches the government has been using is targeting women for interventions that increase their access to resources such as credit and livestock. Yet, the underlying attitudinal and cultural constraints still continue to receive little attention in the study area. Thus, such policy measures need to be accompanied by a restructuring of the discriminatory norms and practices that are embedded in the institutions of family, gender roles and harmful traditional practices. These norms and practices shape the access to resources of men and women and consequently their ability to cope with shocks and ensure the food security of their households.

Eliminating discriminatory norms and values and societal orders is not something that can be accomplished overnight. However, policy interventions can speed up the process. Under the existing situation, government organs responsible for implementing gender policies exist only in a fragmented state at district level and are almost absent at the lowest administration level of the *kebele*. The findings in the study area show that the implementation of women's policies seriously suffers from institutional capacity at the lower level of administration. Women's affairs offices in government organizations are neither supplied with the required human and material resources, nor are they given equal consideration with other departments. Thus, stretching functional WAO structures to the grassroots level such as *kebeles* should be given serious consideration.

It is obvious that the major agents of change in realizing an equal society are women themselves. Yet, given the high level of internalization of oppressive norms, women's agency as it is now cannot be sufficient to bring positive change unless women themselves are able to overcome their feelings of inferiority and subordination. Thus, extensive gender education is required to change their psychological orientation, boost their confidence and sensitize them to their constitutional rights. Thus, it should be the task of the policy-implementing body, through women's affairs offices and NGOs, to facilitate conditions for rural women to enhance their capacity through education and training. This would be a step forward in the process of eradicating gender-biased beliefs and practices. Informal education about the value of gender equality should also target the family as it is the central nucleus where change can be initiated, nurtured and multiplied.

11.3 The relevance of the sustainable livelihoods framework (SLF) and the "theory of structuration" to the empirical investigation

Applying SLF to the study context

The SLF has provided methodological, organizational and analytical guidance for the empirical investigation in this thesis. The thesis has benefited from the "actor-centred" research approach emphasized in the SLF, in that it encouraged the application of a range of qualitative inquiry tools such as focus group discussions, key informant interviews, participant observation and case stories. The various components of the SLF helped to organize the variables to be included in the questionnaires and thus created favourable conditions for gathering a wide range of information. It was possible to enquire into various aspects of the livelihoods of the local people, such as resources, livelihood and coping strategies and vulnerability contexts, including various physical, social and economic factors that influence the livelihood decisions of rural male- and female-headed households. Thus, the flows and interactions shown in the framework have been systematically adopted to structure the thesis.

The study clearly demonstrated that rural FHHs in the study area face multi-dimensional and complex problems that are of an economic, social, political and cultural nature. All these compound to frame their vulnerability context, against which their food insecurity and coping strategies are measured. From the empirical findings it can be seen that coping strategies are manifestations of actor's agency shaped by the constant interaction of men and women with the existing resources as well as formal/informal institutions. Resources are diverse and apart from allocative resources, authoritative resources such as informal social ties are essential pillars of a rural household's livelihood. Widely accepted informal institutions including norms, perceptions, values and customary practices shape actors' entitlement to resources.

Overall, the SLF provided key insights into the vulnerability of men and women. In other words, it shaped my thinking in that dealing with rural livelihoods requires an understanding of the relationship people have with key institutions which guide their entitlements to resources, as well as constraining or promoting decisions and actions which shape livelihood outcomes such as coping mechanisms and food security. Nevertheless, the SLF by itself was found inadequate to explain gender power relations at the depth required in this thesis. The "theory of structuration" was therefore adopted so as to explicitly deal with gender power relations through the agency-structure relationship.

The "theory of structuration"

The "theory of structuration" has made a conceptual contribution to this study in that it provided theoretical understanding of how men and women farmers' actions are being shaped by the environmental, political social and economic transformations happening elsewhere. Giddens' notion of "the duality of structure" is used to guide the discussion on the interplay between actors' agency and institutions in the social field. This has facilitated the understanding of the link between unequal gender power relations and vulnerability, a key concept used in this thesis to investigate food security and coping mechanisms of households.

Against the above backdrop, the result of this empirical study offers the following three perspectives in light of the "theory of structuration".

Firstly, it was my presumption that inter- and intra-household gender power relations function in two different modes, and female headed households are related to the former than the later. Yet, the study findings showed both function in a more or less non-exclusive manner. A great similarity in perception and decision-making behaviour of female heads which very much resembles to that of traditional "housewives" can be good evidence to this. The reproduction of such structures containing unequal gender power relations even at the absence of male spouses in female headed households shows the overlapping nature of intra- and inter-household gender power relation in the study context. In this regard, the study proved that the agency/decision making of female heads of households is largely informed by the existing gender-based informal institutional structures.

Secondly, Giddens' (1979, 24; 1984, 49) theoretical conceptualization of human action as a matter of "practical consciousness" and "reflexive discursive dialogue" is well substantiated by the coping mechanisms and coping strategies employed by both female- and male-headed households in the study area. In this respect, the study offers an insight concerning the way gender power relation can be understood in such study contexts. Firstly, the study shows that women, regardless of their subordinate power position to men, they proactively plan activities to cope with shocks and stresses, engage in bargaining and negotiations outside their homes, including in their village and in the market place, and at times act by not reacting to certain livelihood stresses. In all of these, they exercise decisions implying that their actions in the face increasing environmental challenges can lead to shifts in the existing gender power relations in the long run.

Thirdly, the findings reflect a perspective that gender-based inequalities with regard to female-headed households should be investigated with due consideration of temporal and spatial dimensions of livelihoods. This is because the daily experiences of men and women are continuously shaped and reshaped through "*their movement in time and space*". The changing perceptions in respect of gender equality among farmers over time, on-farm activities that follow a specific agricultural calendar and off-farm activities that have a seasonal character are some of the factors that show the importance of time in the analysis of livelihoods. Moreover, women's claim to land within a patrilocal residential system, economic benefits accrued from market exchanges at different locations, bargaining and negotiations during various informal reciprocal resource-sharing arrangements within villages are good examples showing the importance of space in social science research. However, this study is limited in its ability to operationalize the analysis of time and space at the required depth needed.

References

Addis, T. (2000) The missing link between micro- and macro-level gender disaggregated economic data in economic policy formulation and planning in Ethiopia. *CERTWID Informs* 6(1), 14-18.

Addis, T., Tesfaye, T., Mwangi, W. and Vekuijl, H. (2001) Gender differentials in agricultural production and decision making among smallholders in Ada, Lume, and Gimbichu *Woredas* of the central highlands of Ethiopia. Mexico, D.F: International Maize and Wheat Improvement Center (CIMMYT) and Ethiopian Agricultural Research Organization (EARO). Mexico: CIMMYT.

Adger, W.N., Kelly, P.M., Winkels, A., Huy, L.Q. and Locke, C. (2002) Migration, remittance, livelihood trajectories, and social resilience. *Ambio* 31(4), 358-366.

Agarwal, B. (1986) Women, poverty and agricultural growth in India. *The Journal of Peasant Studies* 14(4), 165-220.

Agarwal, B. (1988) Who sows? Who reaps? Women and land rights in India. *Journal of Peasant Studies* 15(4), 531- 81.

Agarwal, B. (1994) A field of one's own: gender and land rights in South Asia. Cambridge: Cambridge University Press.

Agarwal, B. (1997) Bargaining and gender relations: within and beyond the household. *Feminist Economics* 3(1), 1-51.

Alemayehu, M. (2006) Country pasture/forage resource profiles, Ethiopia. Rome, Italy: FAO.

Ali, H. (2000) Female-headed household vulnerability and their participation in EGS: a case study of two peasant associations in Mekdela *Wereda*, South Wello. MA thesis. School of Graduate Studies, Addis Ababa University.

Almaz, N., and Anke, N. (2004) The significance of enset culture and biodiversity for rural household food and livelihood security in southwestern Ethiopia. *Agriculture and Human Values* 21(1), 61-71.

Appendini, K. (2001) Land and livelihood: what do we know, and what are the issues? In: Zoomers A. (ed): Land and livelihood in Latin America: what do we know, and what are the issues? Amsterdam: Royal Tropical Institute/Vervuert Verlag, 23-38.

Aregu, L., Puskur, R. and Bishop- Sambrook, C. (2011) The role of gender in crop value chains in Ethiopia. A presentation prepared for the gender and market oriented agriculture (AgriGender 2011) workshop, Addis Ababa, Ethiopia, 31st January-2nd February 2011. Nairobi, Kenya: International Livestock Research Institute (ILRI).

Ashley, C. and Carney, D. (1999) Sustainable livelihoods: lessons from early experience. London: Department for International Development (DFID).

Ashley C. and Maxwell, S. (2001) Rethinking rural development. *Development Policy Review* 19(4), 395-425.

Aster, B. and Azeb, B. (2004) Assessment of the impact of gender based violence in Meskan, Mareko and Soddo *woredas* of SNNPRS. Unpublished document. Progynist, Addis Ababa.

Auerbach, C. F. and Silverstein, L. B. (2003) Qualitative data: an introduction to coding and analysis. New York and London: New York University Press.

Ayele, S., Assegid, W., Jabbar, M. A., Ahmed, M. M. and Belachew, H. (2003) Livestock marketing in Ethiopia: A review of structure, performance and development initiatives. Socio-economics and policy research working paper 52. Nairobi, Kenya: International Livestock Research Institute (ILRI).

Bebbington, A. (1999) Capitals and capabilities: a framework for analyzing peasant viability, rural livelihoods and poverty. *World Development* 27(12), 2021-44.

Berger, P. and Luckmann, T. (1967) The social construction of reality. A treatise in the sociology of knowledge. New York: Penguin Books.

Bernard, H. R (2011) Research methods in Anthropology: qualitative and quantitative approaches. Lanham, USA: AltaMira. Press. 5th ed.

Berry, R. A., and Cline, W. R. (1979) Agrarian structure and productivity in developing countries, Baltimore: Johns Hopkins.

Bezabih, E. (2003) Determinants of multiple technology adoption: defining adopters and empirical analysis. *Ethiopia Journal of Agricultural Economics* 5(1-2), 1–22.

Bigsten, A., Kebede, B., Shimeles, A., and Taddesse, M. (2002) Growth and poverty reduction in Ethiopia: Evidence from household panel surveys. *World Development* 31(1), 87.

Birkmann, J. (2011) Regulation and coupling of society and nature in the context of natural hazards. In: Brauch G. H., Spring O.U., Mesjasz C., Grin J., Kameri-Mbote P., Chourou B., Dunay P., Birkmann J. (Eds) Coping with global environmental changes, disasters and security. Threats, challenges, vulnerabilities and risks. Berlin Heidelberg: Springer, 1103-1127.

Bishop-Sambrook, C. (2004) Gender Analysis: An overview of gender issues in the agricultural sector of Ethiopia. IPMS (Improving Productivity and Market Success) gender analysis and strategy paper, Addis Ababa, Ethiopia.

Bohle, H. G., Downing, T. E., Watts, M.J. (1994) Climate change and social vulnerability: toward a sociology and geography of food insecurity. *Global Environmental Change* 4(37), 48.

Boserup, E. (1970) Woman's role in economic development, New York: St. Martin's Press.

Buvinic, M. and Gupta, G. R. (1997) Female-headed households and female-maintained families: are they worth targeting to reduce poverty in developing countries? *Economic Development and Cultural Change* 45(2), 259-280.

Carney, D. (1998) Sustainable rural livelihoods, what contribution can we make? London: DFID.

Carney, J. and Watts, M. (1991) Disciplining women? Rice, mechanization and the evolution of Mandinka gender relations in Senegambia. *Journal of Women in Culture and Society* 16(4), 651-81.

Carswell, G. (2000) Livelihood diversification in southern Ethiopia. IDS Working Paper 117. Brighton: IDS.

Carswell, G. (2002) Livelihood diversification: Increasing in importance or increasingly recognized? Evidence from southern Ethiopia. *Journal of International Development* 14(6), 789-804.

Chafetz, J. S. (1991) The gender division of labour and the reproduction of female disadvantage: toward an integrated theory. In: Blumberg R. L. (ed): gender, family, and economy: the triple overlap. Newbury Park: Sage Publications, Inc., 74-94.

Chambers, R. (1989) Editorial introduction: vulnerability, coping and policy. *IDS Bulletin* 20(2), 1-7.

Chambers, R. and Conway, G. R. (1992) Sustainable livelihoods: practical concepts for the 21st century. IDS Discussion Paper 296. Brighton: IDS.

CHF International (2005) Gurage livelihood and agriculture development program. Final report submitted to OFDA. SNNPR, Ethiopia. http://pdf.usaid.gov/pdf_docs/PDACF737.pdf accessed on 8/8/2012.

Clay, D. C, Daniel, M. and Debebe, H. (1999) Food aid targeting in Ethiopia: a study of who needs it and who gets it. *Food Policy* 24(4), 391-409.

Connelly, M. P., Murray Li, T., MacDolad, M. and Palpart, J. L. (2000) Feminism and development: theoretical perspectives. In: Jane Parpart, L., Patricia, M., Connelly, M.P and Eudine Barriteau, V. (eds): Theoretical perspectives on gender and development. Ottawa, ON, Canada: International Development Research Centre, 51- 159.

Corbett, J. (1988) Famine and household coping strategies. *World development* 16(9), 1099-1112.

Corbetta, P. (2003) Social research: theory, methods and techniques. London, UK: Sage Publications Ltd.

Creswell, J. W. (2003) Research design: qualitative, quantitative, and mixed method approaches. Thousand Oaks, California: Sage Publications. 2nd ed.

Crewett W., Bogale A. and Korf B. (2008) Land tenure in Ethiopia: continuity and change, shifting rulers, and the quest for state control. CAPRI working Paper No. 91. Washington DC USA: ILRI (International Food Policy Research Institute).

Crosbie, T. (2006) Using activity diaries: some methodological lessons. *Journal of Research Practice* 2(1), 1-13.

CSA (2011) Agriculture in figures. Key findings of the 2008/09-2010/11 agricultural sample surveys for all sectors and seasons. Country summary. Addis Ababa: Federal Democratic Republic of Ethiopia (FDRE).

CSA (2012) Ethiopia's rural facilities and services ATLAS 2011. SNNP regional state. Volume 7 http://www.csa.gov.et/images/documents/pdf_files/atlas/SNNPR_RSF.pdf accessed on 8/2/12

CSA (Ethiopia) and ICF International (2012) Ethiopia demographic and health survey 2011. Addis Ababa, Ethiopia and Calverton, Maryland, USA: Central Statistical Agency and ICF international.

CSA (Ethiopia) and ORC Macro (2001) Ethiopia demographic and health survey 2000. Addis Ababa, Ethiopia and Calverton, Maryland, USA: Central Statistical Authority and ORC Macro.

Dasgupta, S. and Maiti, K. A. (1986) The rural energy crisis, poverty and women's roles in five Indian villages', world employment programme technical co-operation report. Geneva: International Labour Office.

Davies, S. (1993) Are coping strategies a cope out? *IDS Bulletin* 24(4), 60-72.

Davies, S. (1996) Adaptable livelihoods: coping with food insecurity in the Malian Sahel. London: Macmillan Press Ltd.

de Haan, L. and Zoomers, A. (2005) Exploring the frontier of livelihoods research. *Development and Change* 36(1), 27-47.

de Janvry and Sadoulet, E. (2001) Access to land and land policy reforms. *Policy Brief No.* 3. Helsinki, Finland: the United Nations University, World Institute for Development Economics Research.

De Waal, A. (1987) Famines that kills: Dafur Sudan, 1984-85. Oxford: Clarendon.

Degefa, T. (2005) Rural livelihoods, poverty and food security in Ethiopia. A case of study at Erenssa and Garbi, Amhara National Regional State. Published Doctoral Thesis, Norwegian University of Science and Technology. Faculty of Social Sciences and Technology Management, Trondheim.

Degnet, A., Belay K. and Aregay, W. (2003) Adoption of high-yielding maize varieties in Jimma zone: evidence from farm level data. *Ethiopian Journal of Agricultural Economics* 5 (1–2), 41–62.

Dejene, A. (1994) Female-headed farm households in two contrasting regions in Ethiopia: access to and management of resources. *Ethiopian Journal of Development Research* 16 (1), 1-13.

Dercon, S. (1996) Risk crop choice, and savings: evidence from Tanzania. *Economic Development and Cultural Change* 44(3), 385-514.

Dercon, S. (2000) Income risk, coping strategies and safety nets. Background paper world development report 2000/01. WPS/2000.26. Oxford University.

Dercon, S. and Krishnan, P. (1996) Income portfolios in rural Ethiopia and Tanzania: choices and constraints. *Journal of Development Studies* 32(6), 850-875.

Dessalegn, R. (1991) Famine and survival: a case study from north east Ethiopia. Sweden: Bohusläningens Boktryckeri AB, Uddevalla.

Dessalegn, R. (1994) Land policy in Ethiopia at the cross roads. In: Dessalegn R. (ed): Land tenure and land policy in Ethiopia after the Derg. Working Papers on Ethiopian Development. Norway: Center for Environment and Development, University of Trondheim, 1-20.

Devereux, S. (2000) Food security in Ethiopia. A discussion paper for DFID, IDS Sussex.

Devereux, S. (2001) Social protection for the poor. IDS working Paper 142. Brighton Sussex: IDS.

Devereux, S. (1993) Theories of famine. New York and London: Harvester Wheatsheaf.

Devereux S., Teshome, A. and Sabates-Wheeler, R. (2005) Too much inequality or too little? inequality and stagnation in Ethiopian agriculture. *IDS Bulletin* 36(2), 121-126.

DFID (1999) Sustainable livelihoods guidance sheets. http://www.ennonline.net/pool/files/ife/dfid-sustainable-livelihoods-guidance-sheet-section1.pdf accessed on 10/11/2012.

DFID (2002) Eliminating Hunger: DFID food security strategy and priorities for action. Consultation document. London: Department for International Development (DFID).

DPED (1998) Gurage zone socio-economic and base- line survey. Wolkite, SNNPR: Department of Planning and Economic Development. Unpublished document.

Drèze, J. and Sen, A. K. (1995) India: economic development and social opportunity. Delhi: Oxford University Press.

Ellis, F. (2000) Rural livelihoods and diversity in developing countries. New York: Oxford University Press.

Ellis, F. and Biggs, S. (2001) Evolving themes in rural development 1950s-2000s. *Development Policy Review* 19(4), 437-448.

Elson, D. (1995) Gender awareness in modeling structural adjustment. *World development* 23(11), 1851-1868

Etenesh, B. (1999) Institutional factors maintain the gender gap in entitlements to rural resources. Paper presented at the fourth conference of the agricultural economics society of Ethiopia, Addis Ababa.

Fafchamps, M. and Quisumbing, A. R. (2000) Control and ownership of assets within rural Ethiopian households. Economics Series Working Papers WPS/2000-27. University of Oxford, Department of Economics.

Fafchamps, M. and Quisumbing, A. R. (2005) Assets at marriage in rural Ethiopia (2005). *Journal of Development Economics* 77(1), 1-25.

FAO (2009) Women and the right to food: international law and state practice. Right to food unit. Rome, Italy: Viale delle Terme di Caracalla.

FAO (2011a) The state of food and agriculture: women in agriculture: closing the gender gap for development. Rome, Italy: Viale della terme di Caracalla.

FAO (2011b) The role of women in agriculture. ESA Working Paper No. 11-02. Rome, Italy: Viale della terme di Caracalla.

Farrington, J., Carney, D., Ashley, C. and Turton, C. (1999) Sustainable livelihoods in practice: early applications of concepts in rural areas. Natural resources perspectives 42. London: Overseas Development Institute.

FDRE (1995) Proclamation No. 1/1995. The Constitution of the Federal Democratic Republic of Ethiopia. *Federal Negarit Gazetta* 1st year No. 1, National legislative bodies, Addis Ababa, Ethiopia.

FDRE (1998) Proclamation No. 147/1998: A proclamation for cooperative societies. *Federal Negarit Gazeta*, National legislative bodies, Addis Ababa, Ethiopia.

FDRE (2000) Proclamation No. 231/2000. The revised family code. *Federal Negarit Gazetta* extra ordinary Issue No. 1/2000, National legislative bodies, Addis Ababa, Ethiopia.

FDRE (2004) Proclamation No. 377/2003. A proclamation for labour. *Federal Negarit Gazetta*. 10th year No. 12, National legislative bodies, Addis Ababa, Ethiopia.

FDRE (2005a). Proclamation No 456/2005. A proclamation for rural land administration. *Federal Negarit Gazetta No.44,* National legislative bodies, Addis Ababa, Ethiopia.

FDRE (2005b): (Ethiopia), Proclamation No.414/2004. The criminal code of the Federal Democratic Republic of Ethiopia. In *Federal Negarit Gazeta* (a separate volume), National legislative bodies, Addis Ababa, Ethiopia.

Fischer, M. (2006) Civil society in conflict transformation: ambivalence, potentials and challenges. Berlin: Berghof Research Centre for Conflict Management.

Flintan, F., Demlie, S., Awol, M., Humed, Z., Belete Y. and Lemma, H. (2008) Study on women's property rights in Afar and Oromiya regions, Ethiopia, USAID and CARE REPORT.

Frank, E. (1999) Gender, agricultural development and food security in Amhara, Ethiopia: the contested identity of women farmers in Ethiopia. Paper prepared for USAID Ethiopia, Addis Ababa.

Gertler, P. and Gruber, J. (2002) Insuring consumption against illness. *American Economic Review* 92(1), 51-70.

Getachew, D. (1995) Economy at the crossroad: famine and food security in rural Ethiopia. Addis Ababa: Care International in Ethiopia.

Giddens, A. (1979) Central problems in social theory. London, UK: Macmillan.

Giddens, A. (1984) The constitution of society: outline of the theory of structuration. Cambridge: Polity Press.

Giddens, A. (1993) Sociology. Cambridge: Polity Press.

Giddens, A (1997) Die Konstitution der Gesellschaft. Grundzüge einer Theorie der Strukturierung. Frankfurt a. Main, New York, 3rd Edition.

Goudie, A. (1998) Eliminating world poverty: a challenge for the 21st century. An overview of the 1997 white paper on international development. *Journal of International Development* 10 (2), 167-183.

Greene, J. C., Caracelli, V., and Graham, W. F. (1989) Toward a conceptual framework for mixed- method evaluation designs. *Educational Evaluation and Policy Analysis* 11(3), 255-274.

Hallward-Driemeier, M. and Gajigo, O. (2013) Strengthening economic rights and women's occupational choice: the impact of reforming Ethiopia's family law. Policy Research Working Paper 6695. Washington DC: The World Bank.

Hancock B., Windridge, K. and Ockleford, E. (2007) An introduction to qualitative research. Nottingham and Scheffield: The National Institute for Health Research RDS for East Midlands/Yorkshire and Humber.

Haregewoin, C. and Emebet, M. (2002) Country gender profile Ethiopia. Addis Ababa: Ethiopia SIDA (Swedish International Development Cooperation Agency).

Haregewoin, C. and Emebet, M. (2003) Towards gender equality in Ethiopia: profile of gender relations. Stockholm: Sweden. SIDA (Swedish International Development Cooperation Agency).

Harriss, B. (1990) The intra family distribution of hunger in South Asia. In: Dreze, J. and Sen A.K. (eds): The political economy of hunger. Clarendon Press: Oxford, 351-424.

Heijmans, A. (2001) Vulnerability: a matter of perception. Disaster Management Working Paper 4/2001. University College of London: Benfield Greig Hazard Research Centre.

Helmke, G. and Levitsky, S. (2004) Informal institutions and comparative politics: a research agenda. *Perspectives on Politics* 2(04), 725-740.

Holmes, R. and Jones, N. (2009) Gender inequality, risk and vulnerability in the rural economy: re-focusing the public works agenda to take account of economic and social risks. Background report for SOFA 2010. London: Overseas Development Institute.

Hussein, K. and Nelson, J. (1998) Sustainable livelihoods and livelihood diversification. IDS Working Paper 69. Brighton: IDS.

IIRR-Ethiopia (2000) Integrating gender in development projects: a manual for field practitioners. Addis Ababa, Ethiopia: International Institute of Rural Reconstruction.

Kabeer, N. (1999) Resources, agency, achievements: reflections on the measurement of women's empowerment. *Development and Change* 30 (3), 435-464.

Kabeer, N. (2003) Gender mainstreaming in poverty eradication and the millennium development goals: a handbook for policy makers and other stakeholders. Ottawa, Canada: The Common Wealth Secretariat, International Development Research Centre, and the Canadian International Development Agency.

Kelly, P. M. and Adger, W. N. (2000) Theory and practice in assessing vulnerability to climate change and facilitating adaptation. *Climatic Change* 47(4), 325-352.

Krantz, L. (2001) The sustainable livelihood approach to poverty reduction. Stockholm: SIDA (Swedish International Development Cooperation Agency).

Kumar, N. and Quisumbing, A. (2012) Inheritance practices and gender differences in poverty and wellbeing in rural Ethiopia. *Development Policy Review* 30 (5), 573-595.

Lado, C. (1992) Female labour participation in agricultural production and the implications for nutrition and health in rural Africa. *Social Science and Medicine* 34 (7) 789-807.

Lastarria-Cornhiel, S. (1997) Impact of privatization on gender and property rights in Africa. *World Development* 22(10), 1455-75.

Leach, M., Mearns, R. and Scoones, I. (1999) Environmental entitlements: dynamics and institutions in community-based natural resource management. *World Development* 27(2), 225-47.

Legesse, D. (2003) Empirical analysis of duration of herbicide adoption in tef-based farming system of west Shewa Zone. *Ethiopian Journal of Agricultural Economics* 5(1–2), 1–22.

Lele, S. M. (1991) Sustainable development: a critical review. *World Development* 19 (6), 607-621.

Lipton, M. and Longhurst, R. (1989) New seeds and poor people. London: Unwin Hyman.

Long, N. (1997) Agency and constraint, perceptions and practices. A theoretical position. In: H. de Haan and N. Long(eds): Images and realities of rural life. Assen: van Gorcum, 1-20.

Mack, N., Woodsong, C., Macqueen Kathleen, M., Guest, G. and Namey, E. (2005) Qualitative research methods: a data collector's field guide. USA: Family Health International.

Mesfin, W. (1984) Rural vulnerability to famine in Ethiopia, 1958-1977. New Delhi: Vikas Publishing House.

Meskan District Agriculture and Development Office (2011) Productive safety net program, plan for 2011 budget year. Unpublished document. Butajira.

Meskan District Finance and Economic Development Office (2011) Unpublished hand out. Meskan district development data collection and distribution office. Meskan, SNNPR.

Meskan District Rural Development Office (2008) District level summary report. Unpublished document. Butajira.

Millman, S. and Kates, R. W. (1990) Toward understanding hunger. In: Newman, L. F, Crossgrove, W., kates, R.W., Mathews, R. and Millman, S (eds): Hunger in history: food shortage, poverty, and deprivation. Cambridge: Basil Blackwell, 3-24.

Mirutse D., Gebregiorgis, H. and Selam, A. (2006) Female-headed households and livelihood interventions in four selected Weredas in Tigray, Ethiopia. Report (44). Oslo: Drylands Coordination Group.

MoARD (2006) Productive safety net programme: programme implementation manual version 2. Addis Ababa, Ministry of Agriculture and Rural Development, FDRE.

MoARD (2009a) Food security program draft paper 2010-2014. Addis Ababa: FDRE.

MoARD (2009b) Productive safety net programme component document 2010-2014. Addis Ababa, FDRE.

MoARD (2009c) Food security program draft paper 2010-2014. Household asset building component. Addis Ababa, FDRE.

MoARD (2010) An atlas of Ethiopian livelihoods. The livelihood integration unit, USAID and FDRE.

MoFED (2006) Ethiopia: building on progress. A Plan for Accelerated and Sustained Development to End Poverty (PASDEP) (2005/06-2009/10). Volume II Policy Matrix. Addis Ababa, FDRE.

MoFED (2010) Growth and Transformation Plan 2010/11-2014/15. Volume one main text. Addis Ababa, FDRE.

Mogues, T., Cohen, M. J, Birner, R., Lemma, M., Randriamamonjy, J., Tadesse, F. and Paulos, Z. (2009) Agricultural extension in Ethiopia through a gender and governance lens. Ethiopia Strategy Support Program 2 (ESSP2) Discussion Paper 007. Addis Ababa: IFPRI (International Policy Research Institute).

Molyneux, M. (1985) Mobilization without emancipation: women's interests, state and revolution in Nicaragua. *Feminist Studies* 11 (2), 227-254.

Morduch, J. (1990) Risk, production, and saving: theory and evidence from Indian households. Harvard University. Department of Economics, Cambridge: Mass.

Moser, C. (1993) Gender planning and development: theory, practice and training. London: Routledge.

Moser, C. (1996) Confronting crisis. A summary of household responses to poverty and vulnerability in four poor urban communities. Environmentally sustainable development monograph series No. 7. Washington DC: The World Bank.

Moser, C. (1998) The asset vulnerability framework: reassessing urban poverty reduction strategies. *World Development* 26(1), 1-19.

Nooteboom, G. (2003) A matter of style: social security and livelihood in upland east Java. Amsterdam: Rozenberg.

Odame, H., Hafkin, N., Wesseler, G. and Boto, I. (2002) Gender and agriculture in the information society. International Service for National Agriculture Research. Briefing Paper No.55. The Haugue, The Netherlands: ISNAR.

ODI (1998) The UK white paper on international development- and beyond. Briefing Paper. Overseas Development Institute 1998. ISSN 0140-8682. http://www.odi.org.uk/sites/odi.org.uk/files/odi-assets/publications-opinion-files/2622.pdf accessed on 6/52013.

Ogato, G.S., Boon E.K. and Subramani, J. (2009) Improving access to productive resource and agricultural services through gender empowerment: a case study of three rural communities in Ambo district, Ethiopia. *Journal of Human Ecology* 27 (2), 85-100.

Progynist (2002) A three year action plan (2002-2005) to carry out an integrated development in the rural and semi-urban setting of Meskan and Mareko Woreda, Gurage zone, unpublished document. SNNPRS, Addis Ababa.

Quisumbing, A., Hadded, L. and Pena, C. (1995) Gender and poverty: new evidence from 10 developing countries. Food consumption and nutrition division. Discussion Paper No. 9. Washington, DC: IFPRI.

Quisumbing, A. and Yohannes, Y. (2005) How fair is workfare? Gender, public works and employment in rural Ethiopia. Policy Research Working Papers 3492. Washington DC: World Bank.

Razivi, S. (1997) From rages to riches. Looking at poverty from a gender perspective. *IDS Bulletin* 28 (3), 49-62.

Rebeka, A. (2006) Impact assessment of rainwater harvesting ponds: the case of Alaba Woreda, Ethiopia. MSC thesis. Natural resource and environmental economics, Addis Ababa University. Addis Ababa, Ethiopia.

Saito, K. A. and Weidemann, C. J (1990) Agricultural extension for women farmers in Africa. Policy, research and external affairs working papers. Washington, DC: The World Bank.

Sara, W. and Muluneh, W. (2013) Livelihood strategies of rural women with emphasis on income diversification and demographic adjustment in central Ethiopia: the case of Olonkomoi, Oromia Region. *The Ethiopian Journal of the Social Sciences and Humanities* 9(1), 111-146.

Schultz, T. W. (1964) Transforming traditional agriculture. New Haven: Yale University Press.

Scoones, I. (1998) Sustainable rural livelihoods: a framework for analysis. IDS Working Paper No. 72. University of Sussex.

Scott, R. W. (2008) Institutions and organizations. Ideas and interests. London, New Delhi, Singapore: Thousand Oaks.

Sen, A. K (1981) Poverty and famines: An essay on entitlement and deprivation. Oxford: Clarendan Press.

Serrat, O. (2010) The sustainable livelihood approach. Washington DC: Asian Development Bank (ADB).

Skapa, B. (1988) When the brain is sleepy, you also sleep: women and agriculture in Zimbabwe. A report to World Bank, Washington DC.

Sosena, D. (2000) Historical background about gender and the development of policy approach to women's involvement in development. Paper presented for training on gender and development program, CRDA, Addis Ababa.

Sosena, D. and Tsahai, Y. (2008) A Review of national policy on Ethiopian women (2008). In: Taye. A. (ed): Digest of Ethiopia's national policies, strategies and programs. Addis Ababa: Forum for Social Studies (FSS), 93-126.

Spring, A. (1985) The women in agricultural development project in Malawi: making gender free development work. In: Gallin R. S. and Spring A. (Eds). Women creating wealth: transforming economic development. Association for women in development: Washington DC, 71-75.

Teklu, T. (2003) Rural lands and evolving tenure arrangements in Ethiopia: Issues, evidence and policies. FSS Discussion Paper No. 10. Forum for Social Studies, Addis Ababa.

Teklu, T. (2005) Land scarcity, tenure change and public policy in the African case of Ethiopia: Evidence on efficacy and unmet demands for land rights. International conference on African development archives. Paper 89.
http://scholarworks.wmich.edu/africancenter_icad_archive/89

Tesfaye, T. and Adugna, L. (2004) Factors affecting entry and intensity in informal rental land markets in southern Ethiopian highlands. *International Agricultural Economics* 30 (2), 117-128.

Tesfaye, T. and Degefa, T. (2009) Linkages between water supply and sanitation (WSS) and food security in four case study communities of East Hararghe zone, Oromia Region. *Ethiopian Journal of Development Research* 31(1), 63-92.

Tesfaye, T. (2012) Water infrastructure and food security linkages in three selected regions of Ethiopia. *Eastern Africa Social Science Research Review (EASSRR)* 28(1), 27-57.

The New Coalition for Food Security (2003) Resettlement as a response to food insecurity: the case of Southern Nations, Nationalities, and People's Region. NGO Discussion Paper. Assessment mission; 12 May-02 June 2003. Addis Ababa.

Tröger, S. (2002) Sustainability in the context of natural and social environments. An actor oriented interpretation from South west Tanzania. *ERDKUNDE* 56(2), 170-183.

Tröger, S., Grenzebach, H., Heilde, Z. F., Kuhnert, G., Lang, B., Peth, S., Pfitzner, J., Sefrin, C. and Thiele, A. (2011) Failing seasons, ailing societies: climate change and meaning of adaption in Ethiopia. Addis Ababa: Heinrich Böll Foundation Ethiopia.

UN (1993) Declaration of elimination of violence against women 48/104. General assembly http://www.un.org/documents/ga/res/48/a48r104.htm accessed on 6/5/2013.

UN (2012) International Covenant on Economic and Social and Cultural Rights (ICESCR). Office of Higher Commissioner for Human Rights. http://www.ohchr.org/EN/ProfessionalInterest/Pages/CESCR.aspx accessed on 6/5/2013.

UN ECOSOC (1997) Report of the economic and social council for 1997. A/52/3. http://www.un.org/documents/ga/docs/52/plenary/a52-3.htm accessed on 13/12/2013.

USAID (2005) SNNPR livelihood zone reports: SNNPR Follow-on to regional livelihoods baseline study. Chemonics International Inc.

WAO and WB (1998) Implementing the Ethiopian national policy for women, institutional and regulatory issues. Washington DC: The International Bank for Reconstruction and Development/ The World Bank.

Watts, M. (1983) Silent violence: food, famine and peasantry in Northern Nigeria. Berkeley: university of California Press.

Watts, M. J. and Bohle, H. G. (1993) The space of vulnerability: the causal structure of hunger and famine. *Geo Journal* 30 (2), 43-67.

Webb, P. and Von Braun, J. (1994) Famine and food security in Ethiopia. Lessons for Africa. Chichester: John Wiley Ltd.

Whitehead, A. and Lockwood, M. (1999) Gender in the World Bank's poverty assessments: six case studies from Sub-saharan Africa, Discussion Paper No. 99. United Nations Research Institute for Social Development.

Wisner, B., Blaikie, P., Cannon, T. and Davis, I. (1994) At risk: natural hazards, peoples vulnerability and disasters. London: Routledge.

Workneh, N. (2008) Food security strategy and productive safety net program in Ethiopia. In: Taye A. (ed): Digest of Ethiopia's national policies, strategies and programs. Addis Ababa: Forum for Social Studies (FSS), 1-22.

Worku, N. (1996) Gurage Urban migration and the dynamics of cultural life in the village. In: Hudson, G. (ed): Essays on Gurage language and culture. Wiesbaden: Otto Harrassowitz Verlag, 133-151.

World Bank (1989) Kenya: the role of women in economic development. World Bank country studv. Washington, DC: The World Bank. http://documents.worldbank.org/curated/en/1989/05/437471/kenya-role-women-economic-development accessed on 10/10/2013.

World Bank (2013) Working for a world free of poverty. http://data.worldbank.org/country/ethiopia accessed on 2/10/2013 accessed on October 2, 2013.

Yared, A. (1995) Land distribution and its implications for peasant differentiation in Wogda, Northern Shewa. *In Ethiopian Journal of Development Research* 17(1), 1-22.

Yared, A. (1997) Women's access to resources in Amhara households of Wegda, Northern Shewa. In: Fukui K., Kurimoto E., and Shigeta M. (eds): Ethiopia in Broader Perspective. Papers of the 13th International Conference of Ethiopian Studies, Vol. III. Kyoto, 747-758.

Yigremew, A. (2000) Preliminary assessment of the impacts of the post Derg land policy on agricultural production: a case study of land redistribution in two communities in northwest Ethiopia. In: Workneh, N., Legesse D., Abebe, H., Solomon B. (eds): Institutions for Rural Development, Proceedings of the 4th Annual Conference of Agricultural Economics Society of Ethiopia. 23-24 Novemeber 1999, Addis Ababa, 173-190.

Yigremew, A. (2001) Land redistribution and female-headed households: a study in two rural communities in northwest Ethiopia. FSS Discussion Paper No. 5, Addis Ababa: Forum for Social Studies (FSS).

Yigremew, A. (2004) Access to land in rural Ethiopia. A desk review. A Study report submitted to sustainable land use Forum. May 2004, Addis Ababa.

Yigremew, A. (2005) Rural women's access to land in Ethiopia. Addis Ababa, Ethiopia: Forum for Social Studies. www.fssethiopia.org.et/rural_women.htm accessed on 8/17/2013.

Yigremew, A. (2006) Rural women's access to land in Ethiopia. In: Dessalegn R. and Taye A. (eds): Land and the challenges of sustainable development in Ethiopia. Addis Ababa: Forum for Social Studies (FSS), 19-40.

Yin, R. K. (2009) Case study research: design and methods. Califonia, USA: Sage Publications, Inc. 4th ed.

Yohannes, K., Kisan, G. and Garrth, C. (1990) Adoption of new technologies in Ethiopian agriculture: the Case of Tegulet- Bulga District, Shoa province. *Agricultural Economics* 4(1), 27-43.

Zenebework T. and Yared, A. (2000) Women's land rights in Ethiopia. *Birchi: The Journal of Ethiopian Women Lawyers Association* 1 (1), 25- 51.

Appendix I

HOUSEHOLD SURVEY QUESTIONS

The aim of this questionnaire is to gather quantitative data for a study entitled Households' Food Insecurity and Coping strategy in the Face of Vulnerability: a Gender Perspective with Empirical Evidence from Meskan District, Southern Ethiopia. The questions are divided into five parts. I would like to thank you in advance for your patience and your willingness to provide genuine and appropriate responses to each question. The responses and opinions you give will be used exclusively for the purpose of this research.

NAME OF ENUMERATOR: ______________________________

DATE OF DATA COLLECTION: ____________________________

NAME OF THE *KEBELE*: ________________________________

CODE NUMBER: ____________________________________

PART I: QUESTIONS RELATED TO THE DEMOGRAPHIC CHARACTERISTICS OF HOUSEHOLDS

1.1 Sex of household head ___________________
1.2 Age of household head ___________________
1.3 Educational level________________________
1.4 Religion ____________________
1.5 Household size ______________
1.6 Marital Status: 1. Single 3. Divorced 5. Separated
2. Married 4. Widower/d
6. Other (specify)_____________
1.7 Occupation/ main activity _____________________
1.8 If you have given "farmer" in response to 1.7, which kind of agricultural activity do you undertake?
1. Crop production 2. Gardening
3. Growing trees 4. Rearing of cattle and small ruminants
5. Poultry 6. Others (specify) ___________________
1.9 Which of the following agricultural activities do you undertake?

Activities	✓ / X
Preparation of land	
Ploughing	
Planting	
Weeding	
Harvesting produce	
Threshing	
Transporting produce	
Storing	
Feeding domestic animals	
Herding domestic animals	
Milking	
Cleaning barn	

1.10 Are there any absent household members?
1. Yes 2. No
1.11 If "yes", relation to the family
1. Husband 2. Wife 3. Son/Daughter 4. Other____________
1.12 If you have answered "yes" to question 1.10, what is the reason for their absence? (Multiple responses are possible)
1. Seasonal migration 2. Education
3. Staying with family elsewhere 4. The desire for economic independence
5. Others (specify) _________________

PART 2: QUESTIONS RELATED TO FEMALE HEADSHIP
(Only for female-headed households)

2.1 How did you become head of household?

1. Divorce — 2. Death of husband
3. Unmarried /single — 4. Seasonal migration of spouse
5. Separation — 6. Others (specify) ____________________

2.2 If you have given "divorce" in answer to question 2. 1, what was the cause of divorce? (Multiple responses are possible)

1. Polygamy
2. Conflict over household's resource use and management.
3. Health-related reasons
4. Other (specify) ______________________________

2.4 If you have given "seasonal migration of the spouses" in answer to question 2.1, do you receive a remittance?

1. Yes — 2. No

2.5 If you have given "separation" in answer to question 2.1, what was the cause of the separation?

1. Polygamy — 3. Health issues
2. Poverty — 4. Conflict within the household
5. Other (specify) _________________

PART 3: QUESTIONS RELATED TO HOUSEHOLD'S OWNERSHIP OF AND ACCESS TO LIVELIHOOD ASSETS.

A. Ownership of and access to land

3.1 Do you own farm land? 1. Yes 2. No

3.2 If you have answered "yes" to question 3.1, how many plots of land do you own?

1. One — 2. Two — 3. Three — 4. Four

3.3 If you have answered "yes" to question 3.1, what is the source of your land ownership?

1. The state — 2. Inheritance from parents and relatives
3. Lease — 4. Exchange
5. Rent — 6. Purchase
7. Share cropping
8. Other (specify) ____________________________

3.3 If you have answered "yes" to question 3.1, how large is your farm land (in *timad*)?

____

3.4 How do you rate your farm land in terms of its soil fertility?

1. Fertile — 2. Moderately fertile — 3. Not fertile
4. Other_____________________________________

3.5 If you have answered "no" to question 3.1, what is the reason for not possessing land?

1. No inheritance from parents and relatives
2. Did not receive from the state
3. Have lost it due to divorce
4. Have sold it
5. Other (specify) ____________________________

B. Questions related to ownership of and access to oxen

3.6 Do you own livestock?

1. Yes — 2. No

3.7 If you have answered "yes", please provide the following information about the number of livestock you own and those you have sold in the last five years.

Category	Type of livestock	No of livestock owned	Number of livestock sold during the last 5 years
Cattle	Oxen		
	Cows		
	Calves		
	Heifers		
Pack animals	Donkeys		
	Mules		
	Horses		
Small ruminants	Sheep		
	Goats		
	Poultry		
	Bees		

3.8 How did you acquire the livestock? (Multiple responses are possible)
1. Inheritance 2. Purchase
3. Gift 4. Exchange
5. Through productive safety net program
6. Other (specify) ____________________________

3.9 If you have sold livestock over the last five years, why did you sell?
1. To purchase food 2. To purchase farm animals
3. To pay debts 4. To pay tax and other social obligations
5. To buy clothes 6. Other (specify) ____________________

3.10 Do you often face problems associated with livestock rearing?
1. Yes 2. No

3.11 If your response to 3.10 is "yes", identify the major problems (multiple responses are possible).
1. Shortage of fodder 2. Shortage of grazing land
3. Shortage of water 4. Livestock diseases
5. Lack of knowhow 6. Others (specify) __________________

3.12 What have you noticed regarding your household's livestock ownership during the last 5 years?
1. It is increasing 2. It is decreasing
3. I haven't seen a significant change 4. Other______________

3.13 Have you ever faced shortage of oxen in the last five years?
1. Yes 2. No

3.14 If you have answered "yes" to 3.13, what alternatives do you use to overcome the problem? (Multiple responses are possible)
1. Exchange oxen 3. Borrow relatives/friends oxen
2. Share crop 4. Mekenajo
5. Other (specify) ____

C. Questions related to access to credit

3.15 Do you have information about the presence of a formal credit service in your *kebele*?
1. Yes 2. No

3.16 If you have answered "yes" to 3.15, are you a beneficiary?
1. Yes 2. No

3.17 If you have answered "yes" to 3.16, please specify the name of the credit institution.

____

3.18 If you have answered "No" to 3.16, what are the reasons for not being a beneficiary? (Multiple responses are possible)
1. Fear of debt 2. Unable to pay collateral
3. Lack of knowhow 4. Lack of interest
5. Other (specify) ___________________

3. 19 Do you have access to village-based informal credit and saving institutions in your *kebele*?
1. Yes 2. No

3. 20 If you have answered "yes" to 3.19, specify which of the following.
1. Idir 2. Equb
3. Individual money lenders 4. Other (specify) ________________

D. Questions related to access to labour

3.21 Have you faced shortage of labour to undertake agricultural activities?
1. Yes 2. No

3.22 If you have answered "yes" to 3.21, which alternatives do you use to overcome the problem? (Multiple responses are possible)
1. Hire people 2. Ask help from relatives or friends
3. Debo 4. Others (specify)________________________

E. Questions related to access to agricultural inputs

3.23 With regard to your use of agricultural inputs in the production year 2010, please fill in the following table.

Type of agricultural input		Have you used ✓ / X	Amount used
Fertilizer	DAP		
	Urea		
	Natural fertilizer		
Improved seed	Maize		
	Teff		
	Pepper		
	Insecticide		
	Pesticide		

3.24 The use of agricultural inputs on your land over the last five years
1. has increased 2. has decreased 3. has remained constant

3.25 If you have answered "has decreased" to 3.24, what are the reasons for this?

____

3.26 Do you know of the presence of a DA in your *kebele*?
1. Yes 2. No

3.27 If you have answered "yes" to 3.26, how often did you have contact with the DA during the last one year?
1. Not at tall 2. Sometimes 2. Often 3. Very often

3.28 If you have answered "no" to 3.26, specify the reason.
1. Limited information 2. Cultural barriers 3. Lack of interest
4. Administration problem 6. Other (specify) ________________________

PART 4: QUESTIONS RELATED TO ON-FARM AND NON- FARM ACTIVITIES AND HOUSEHOLDS' FOOD SECURITY

4.1 What types of crops did you produce in the 2010 production year?
____

4.2 What types of livestock and livestock products did you produce in the 2010 production year?
____

4.3 Which was the major grain you produced in 2010? How much did you produce?
____

4.4 Can you specify which products are used for daily consumption, for sale and for both?
Daily consumption____
For sale____
For both____

4.5 Can you please fill out the following table to show the number of months that your crop production meets your household's food requirement under three different rainfall conditions?

Rainfall condition	Number of months production meets household's food demand			
	0-3	4-6	7-9	10-12
Good rainfall				
Medium rainfall				
Poor/bad rainfall				

4.6 How would you describe the maize harvest in 2010 as compared with 2009?
1. Has increased 2. Has decreased 3. Has remained constant

4.7 If you response to 4.6 is "has decreased", can you specify the reason?
____

4.8 Identify the type of non-agricultural activities undertaken in your household (multiple responses are possible)

Types of Activities	**Yes /No** ✓ /X
Wage labour in towns	
Safety net	
Wage labour on other people's farm land	
Petty trading of grain	
Petty trading of fuel wood	
Sale of traditional drinks	
No non-agricultural activities	
Other (specify)	

4.9 How would you describe your household food security status?
1. Self-sufficient 3. Seasonally food insecure
2. Food insecure throughout the year 4. Other____________________

PART 5: QUESTIONS RELATED TO SHOCKS AND COPING STRATEGIES OF HOUSEHOLDS

5.1 Have you experienced livelihood shocks/ crises in the last one year?

1. Yes	2. No

5.2 If your response to 5.1. is "yes", can you specify which kind of livelihood shock/ crisis your household experienced? (Multiple responses are possible)

1. Illness of a family member	2. Death of a family member
3. Death of livestock	4. Crop failure
5. Shortage of income
6. Others (specify) ______________________________

5.3 How many times have you experienced livelihood shocks/ crises in the last five years?

1. Once	2. Twice
3. Three times	4. Four times
5. ≥ five times	6. No shock/crises

5.4 How many times have you experienced livelihood shock/ crises in the last one year?

1. Once	2. Twice
3. Three times	4. Four times
5. ≥ five times	6. No shock/crises

5.5 Can you obtain 100 *birr* to meet an emergency?

1. Yes	2. No

5.6 Which alternatives do you have if you need to obtain 100 *birr*?

1. Selling livestock	2. Selling grain
3. Selling firewood	4. Selling private property
5. Ekub	6. Idir
7. Borrowing from relatives	8. Asking for gifts from friends and relatives
9. Own savings	10. Wage labour
11. Other (specify) ______________________________

5. How does your household cope with food shortage?

____

____

____

Appendix II

Guiding questions for focus group discussions (FGDs)[1]

Part I With separate groups of female- and male-headed households (first round)

1. How would you describe rural life as a female head? (only for female-headed households)
2. What are the most widely practised agricultural activities and non-agricultural activities in your *kebele*?
   a) Who has responsibility for each of these, women, men or both? How are they practised?
   b) How important is each practice in terms of its benefit to your household?
   c) Who decides on their use? How does decision-making take place?
   d) How is cash used? Who decides on cash use? How does decision-making takes place at household level?
3. How would you describe
   a) a well-off household in your *kebele*
   b) a poor household in your *kebele*
   c) the poorest of the poor in your *kebele*
4. Describe the agricultural calendar that is followed in your study *kebele*. Have you perceived any changes over the years?
5. What is the ordinary daily routine for a woman in a female- and a male-headed household? Have you perceived any changes over the years in terms of allocation of time?
6. What is your perception of a food-insecure household? What do you think are the causes of food insecurity?
7. How do you cope with food shortages? Have you noticed any changes in the way your household is coping over the years? What do you think are the reasons? What are the major constraints to effectively coping at household level?
8. What is your opinion about food security programmes delivered by the government in your *kebele*?

Part II Guiding questions for focus group discussions (FGDs) with mixed groups of female- and male-headed households (second round)

A. *Questions relating to informal organizations in the study kebeles*

1. What are the major informal organizations/ local groups currently functioning in the *kebele*? Who participates in them? e.g elders, women. What are their purposes?

   a) Are there local groups organized around environmental issues? e.g. forest users group, water users group. Are leadership positions dominated by a particular social group (high caste, wealthy elderly men, women)? To what extent do women in general and female-headed households in particular enjoy leadership positions and decision-making in these institutions?
   b) Are there local groups organized around economic issues? e.g. credit, agriculture production. Are leadership positions dominated by a particular social group (high caste, wealthy elderly men)? To what extent do women in general and female-headed

[1] The questions were simply used as guides during the discussions. And it has to be underlined that the discussions benefited a lot from probing and flexibility in using the guiding questions.

households in particular enjoy leadership positions and decision-making in these institutions?

c) Are there local groups organized around social issues? e.g. health, literacy, religion. Are leadership positions dominated by a particular social group (high caste, wealthy elderly men)? To what extent do women in general and female-headed households in particular enjoy leadership positions and decision-making in these institutions?

2. Are there any local groups from which women/female-headed households are excluded? Which ones? Why? What do these groups lose due to their lack of participation?
3. Are there groups exclusively for women? Female-headed households? If so, what is the role of these groups?
4. Which informal organizations have so far succeeded in meeting community needs? Example: improving livelihood and food security and coping with shocks in both female- and male-headed households.
5. Which informal institutions have links with formal institutions? For example, NGOs, political parties, government institutions. How do these relations impact the livelihood of the poor and vulnerable rural households such as female-headed households in the *kebeles*?

B. Questions relating to households' perception of food security and vulnerability

1. What is your perception of a food-insecure rural household? Can you identify specific groups of households that you characterize as food insecure in your *kebele*? Are female- and male-headed households equally food insecure? Who is more food insecure? Why?
2. What is your understanding of a vulnerable livelihood? Who is vulnerable in your *kebele*? Are female- and male-headed households equally vulnerable to livelihood stress and shocks? Who is more vulnerable? Why?

Appendix III

List of Key informants

No.	Name	Interview date	Place/ Office	Position/ community role
1	Mr. Wondimagegn Abebe	21/2/2011	Meskan district, Omo Finance and Micro Credit Office	Officer
2	Mr. Endale Kebede	25/2/2011	Dirama *kebele*, Health Extension	Head
3	Mr. Sebro Ibro	24/2/2011	Dirama *kebele*	Shimagile and religious leader
4	Mr. Niguse Gizaw	25/2/2011	Semen Shershera *kebele*, Agricultural Extension	Natural Resource Development, Expert
5	Ms. Rehima Tesema	22/2/2011	Dobo Tuto *kebele*, Agricultural Extension	Livestock Breeding, Expert
6	Mr. Manyazewal Argata		Meskan district, Disaster prevention and food security Desk	Expert
7	Mr. Teshome Mehamed	23/2/2011	Dirama *kebele*, Administration	Head
8	Mr. Zelalem Bekele	23/2/2011	Dirama *kebele*, Agricultural Extension	Livestock Breeding, Expert
9	Ms. Seble Mengesha	24/2/2011	Dirama *kebele*, Saving and Credit Association	Head
10	Mr. Melese Tsegaye	22/2/2011	Beresa *kebele*, Agricultural Extension	Livestock Breeding, Expert
11	Ms. Munteha Redi	21/2/2011	Meskan district, Rural Development and Family Science	Expert for rural women
12	Ms. Amina Mehamed	18/2/2011	Meskan district, Women and Children Affairs Office	Gender Expert
13	Ms. Fantaye Chekol	18/2/2011	Meskan district, Women and Children Affaires Office	Gender and Women's right Expert, volunteer
14	Mr. Kebede Duguma	14/1/2011	Progynist, NGO	Rural program coordinator
15	Ms. Netsanet Mengistu	12/1/2011	Progynist, NGO	Executive director
16	Ms. Lemlem Asfawu	11/1/2011	Progynist, NGO	Gender Expert
17	Mr. Lema Seifu	6/5/2012	Beresa *kebele*	*Shimagile*, *idir* leader, former administration head

Appendix IV

Guiding questions to key informants[2]

A. *Questions relating to physical resources in each* kebele

1. What resources are in ample supply in the *kebele*? Which ones are in short supply? Which are degrading or improving?
2. Who makes decisions about who can use land? Water? Other important resources?
3. Where do you go for water and grazing? How long does it take on average for farm households? How long did it take 5 years ago, 10 years ago?
4. Are the rights of access to these resources different for women and men?
5. Which of the indicated resources are difficult to access for women? Particularly for female-headed households? Why?

B. *Questions relating to public infrastructures and services in each kebele*

1. What public infrastructures and services are available in the *kebele*?
2. What economic opportunities are available in the *kebele*?
3. Is the access to various public infrastructures different for women and men in general and for female- and male-headed households in particular? What are the principal problems?
4. What interventions for improvements have been made?

C. *Questions relating to the demographic and social characteristics of each kebele*

1. How many households are there? What is the total number of people? What is the total number of female- and male-headed households in the *kebele*?
2. Is the village growing or shrinking? Why? (birth-rate, out-migration, in-migration)
3. In what ways do norms and customary practices affect the lives of particular groups such as rural women?
4. What are the local definitions of "rich" and "poor"? Which households are rich? Poor?

D. *Questions relating to environmental, social and economic transformations in each kebele (questions relating to livelihood vulnerability in each kebele)*

1. What are the most important environmental trends over the last 10 years in the locality? e.g. drought, deforestation, erosion
2. What are the most important economic trends over the last 10 years in the locality? e.g. jobs, wages, process, costs of living, crop yields, livestock population.
3. What are the most important demographic trends over the last 10 years? e.g. birth-rates, infant mortality, immigration, out migration, increase in female-headed households
4. What other trends are important in the context of the *kebele*?
5. In your perception, what is getting better? And what is getting worse?
6. Which socio-economic and environmental trends impact women and men differently? What trends affect particularly female-headed households? Why?
7. Which trends impact the poor more than the rich farm households?

[2] The questions were simply used as guides during the interviews held with various key informants. And it has to be underlined that the interviews benefited a lot from probing and flexibility in using the guiding questions.

E. Questions relating to food security and empowerment interventions

1. Which major development and food security programmes are currently being implemented in Meskan district in general and the study *kebeles* in particular? What is the duration of implementation of the programmes? Who are the target groups? Are there any special target beneficiary groups? What are the long term and short term goals and objectives of each programme?
2 To what extent does the Productive Safety Net Programme (PSNP) reach poor rural households? How many *kebeles* have so far been reached? How many are still to be reached? How many households in each *kebele* are able to benefit out of the PSNP? What parameters are used by the government to select beneficiary households?
3. Are there specific programmes which are targeted to benefit female-headed households? What is the impact of the Productive Safety Net Programme with respect to improving the livelihood of the most vulnerable sections of the rural population, such as female-headed households?
4. In principle, the active participation of the local people in the planning, implementation, and the evaluation of local development programmes is a key to ensuring local livelihood sustainability. To what extent is this practiced among the rural agricultural population? Are there any efforts to ensure equal participation of female-headed households in every decision-making process that affects their lives? What are the major constraints to active participation in this regard?
5. How sustainable is the Productive Safety Net Programme in alleviating the food insecurity and vulnerability of rural households in the District? What are its merits and demerits?
6. What are the major women empowerment activities that have taken place so far? Which programmes are currently running in the study *kebeles*?
7. How are gender policies being implemented at district and *kebele* level?
8. What are the major gender-related issues the office has dealt with and is dealing with? What improvements are there? What are the existing gaps?

Appendix V

Types of crops grown in the production year 2010 by the type of household

Crop	Type of Household					
	FHHs		MHHs		Total	
	№	%	№	%	№	%
Wheat	6	3.4	1	0.6	7	2.0
Teff	12	6.9	11	6.1	23	6.5
Maize	52	29.7	32	17.8	84	23.7
Barley	-	-	1	0.6	1	0.3
Beans	-	-	1	0.6	1	0.3
Zengada	-	-	1	0.6	1	0.3
Haricot bean	1	0.6	1	0.6	2	0.6
Teff +maiz	25	14.3	34	18.9	59	16.6
Maize+zengada	16	9.1	9	5.0	25	7.0
Annual crops+ fruits	15	8.6	24	13.3	39	11.0
Zengada +*teff*	11	6.3	9	5.0	20	5.6
Maize+*teff* +wheat	4	2.3	10	5.6	14	3.9
Maize+*teff* + haricotbean	1	0.6	4	2.2	5	1.4
Teff +maize+zengada	6	3.4	8	4.4	14	3.9
Teff+wheat	2	1.1	4	2.2	6	1.7
Maize +zengada + haricot bean	4	2.3	8	4.4	12	3.4
Teff+ zengada+ maize+haricot bean	1	0.6	2	1.1	3	0.8
Maiz +haricot bean	14	8.0	4	2.2	18	5.1
Maize +wheat	3	1.7	1	0.6	4	1.1
Wheat + legumes	2	1.1	-	-	2	0.6
Teff+ maize+beans	-	-	9	5.0	9	2.5
Maize+wheat+ beans+chick pease	-	-	4	2.2	4	1.1
Wheat+ *teff* +chick peas	-	-	1	0.6	1	0.3
Maize+beans+ chick peas	-	-	1	0.6	1	0.3
Total	180	100	175	100	355	100

∑zengada and others in FHHs =38
∑zengada and others in MHHs=36
∑legumes (haricot bean/peas) and others in FHHs=22
∑legumes (haricot bean/peas) and others in MHHs=33

Source: household survey March to April 2011

Appendix VI

Organizational structure at *kebele* level

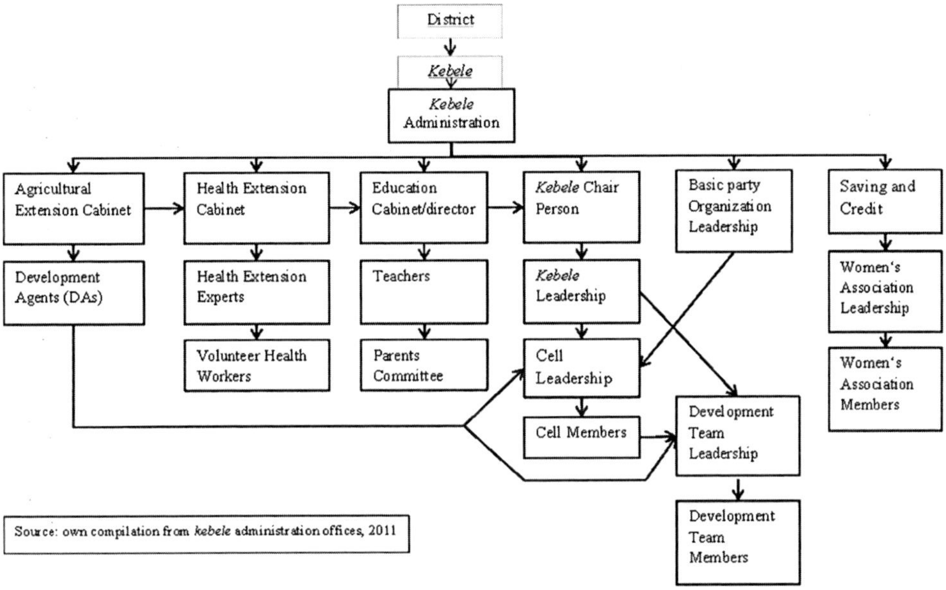

Source: own compilation from *kebele* administration offices, 2011

Spektrum

Berliner Reihe zu Gesellschaft, Wirtschaft und Politik in Entwicklungsländern
hrsg. von Prof. Dr. Dr. Dr. Ulrich van der Heyden, Prof. Dr. Volker Lühr, Prof. Dr. Ulrike Schultz und Prof. Dr. Manfred Schulz

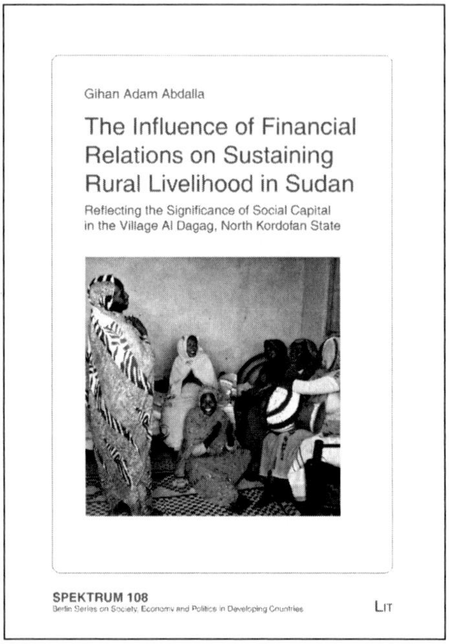

Gihan Adam Abdalla
The Influence of Financial Relations on Sustaining Rural Livelihood in Sudan
Reflecting the Significance of Social Capital in the Village Al Dagag, North Kordofan State

This book started with a vision to portray the Sudanese rural livelihood and the question whether microfinance is able to sustain rural livelihood. Al Dagag village in North Kordofan State, Sudan is chosen to be the research case study. However further questions arose due to using the inductive methodology of Grounded Theory. Throughout the entire research process different categories emerged from the data, which conceptualize the perspective of the rural people on livelihood, microfinance and financial institutions. The path to remodel the livelihood approach led to including new determinants.

Bd. 108, 2013, 240 S., 49,90 €, br.,
ISBN 978-3-643-90403-4

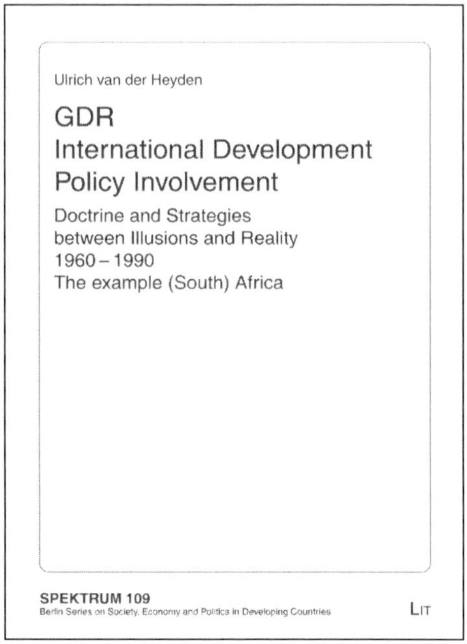

Ulrich van der Heyden
GDR International Development Policy Involvement
Doctrine and Strategies between Illusions and Reality 1960 – 1990. The example (South) Africa

The multi-faceted and far-reaching dealings between the former German Democratic Republic (GDR) and Third World countries, their liberation movements and development policies have not been focussed on before. For almost forty years there has been a disregard of the history and previously intense relationship which was maintained between the GDR and Africa. This situation is arguably incomprehensible as Germany's image in many Third World countries has been defined greatly by the smaller of the two Germanies, especially in South Africa.

Bd. 109, 2013, 320 S., 69,90 €, br.,
ISBN 978-3-643-90421-8

LIT Verlag Berlin – Münster – Wien – Zürich – London
Auslieferung Deutschland / Österreich / Schweiz: siehe Impressumsseite